Samuel Weidman

A Contribution to the Geology of the Pre-Cambrian Igneous Rocks

of the Fox River Valley, Wisconsin

.

Samuel Weidman

A Contribution to the Geology of the Pre-Cambrian Igneous Rocks
of the Fox River Valley, Wisconsin

ISBN/EAN: 9783744743211

Printed in Europe, USA, Canada, Australia, Japan

Cover: Foto ©berggeist007 / pixelio.de

More available books at **www.hansebooks.com**

WISCONSIN GEOLOGICAL AND NATURAL HISTORY SURVEY.

E. A. BIRGE, Director.

BULLETIN NO. III.   SCIENTIFIC SERIES NO. 2.

A CONTRIBUTION

TO THE GEOLOGY OF THE

PRE-CAMBRIAN IGNEOUS ROCKS

OF THE

FOX RIVER VALLEY, WISCONSIN.

BY

SAMUEL WEIDMAN, PH. D.,

Assistant Geologist Wisconsin Geological and Natural History Survey.

I. THE UTLEY METARHYOLITE.
II. THE BERLIN RHYOLITE–GNEISS.
III. THE WAUSHARA GRANITE.

MADISON, WIS.
PUBLISHED BY THE STATE
1898

CONTENTS.

GEOLOGICAL MAP OF THE PRE-CAMBRIAN OUTCROPS OF SOUTHERN WISCONSIN.

IGNEOUS ROCKS OF THE FOX RIVER VALLEY.

INTRODUCTORY.

The areas of igneous rocks to be described in this paper are the outcrops at Utley and Berlin, in Green Lake county, and the several outcrops of granite in Waushara county, in the district of the Fox river. They are a part of the cordon of hills and ridges of pre-Cambrian rocks that lie outside the more continuous crystalline core of central and northern Wisconsin. These outlying hills were islands in the early Paleozoic seas, and around them are deposited the later horizontal formations of sandstone and limestone. Among the pre-Cambrian outliers are to be found rocks of both sedimentary and igneous origin. Those of the Fox river district are igneous, of both plutonic and volcanic nature. They include those at Utley, Berlin, and Waushara, described in this paper, and also those of another number of this bulletin at Montello, Observatory Hill, Marquette, Marcellon, Taylor's farm, and Moundville. Farther to the south and west, at Waterloo, Baraboo, and Necedah, are extensive areas of quartzites. The area of the various pre-Cambrian outliers and their geographic distribution and relation to the pre-Cambrian farther north, are outlined on the general map of Wisconsin. (Pl. II.) On the accompanying map (Pl. I.) is indicated the distribution of these isolated areas on a larger scale.

The outliers vary widely in areal extent and elevation. They indicate that the early Huronian continent had not been reduced to a base-level at the time of the deposition of the early Paleozoic formations. It is very probable that some of the outcrops, at least, rose a thousand feet above the shore line of the early Paleozoic sea.

The outlying pre-Cambrian sedimentary series is best represented in the Baraboo region where there occurs a considerable area of vitreous quartzite elevated into a gentle fold, the remnants of which appear as two parallel ridges about thirty miles in length. In the Baraboo region also, lying above the quartzite along the north flank of the north range of the Baraboo bluffs, is an occurrence of the acid volcanic rock like the rhyolites of the Fox river. The evidence of an erosion interval between the volcanic rock and the underlying quartzite, which is considered to be of the Upper Huronian period, indicates the probable Keweenawan age of the Baraboo volcanics. The similarity in composition of the Baraboo volcanics with the igneous rocks of Fox river, here described, is shown by the following analyses:

	I.	II.	III.	IV.	V.
SiO_2	73.09	73.65	74.62	73.00	71.24
Al_2O_3	13.43	11.19	10.01	15.61	12.20
Fe_2O_3	} 2.57	1.31	3.85	....	1.71
FeO	{	3.25	1.72	1.95	5.44
CaO	2.29	2.78	2.43	.79	.98
MgO	1.03	.51	.33	....	.13
K_2O	1.58	1.86	3.38	.88	1.86
Na_2O	3.85	3.74	3.33	4.95	4.29
MnO	trace	....	....	....	.97
SO_3	....	....	....	.76	....
H_2O	.72	.44	.24	1.06	.81
	98.56	99.23	99.91	99.00	99.63

I. Utley metarhyolite. (Weidman).
II. Berlin rhyolite-gneiss. (Weidman).
III. Waushara granite. (Weidman).
IV. Baraboo keratophyre. (Austin[1]).
V. Baraboo keratophyre. (Daniells[1]).

The close agreement in chemical composition of these rocks is obvious. The amount of silica in the four areas enumerated is practically the same. The relative amount of soda as compared with potash, however, is the distinguishing characteristic. The percentage of soda is much larger than that of potash in all the areas except that of the Waushara granite, and in the latter

---

[1] Bulletin Univ. of Wis., Sci. Ser., No. 2, p. 47.

MAP OF WISCONSIN SHOWING DISTRIBUTION OF THE PRE-CAMBRIAN AND
PALEOZOIC ROCKS AND THE AREA OF GENERAL MAP.

place, if we take into account the molecular ratio, the soda molecule is found to be in excess.

The three areas of the Fox river are similar in comparative amounts of lime, soda, and potash. The Baraboo rock differs from these somewhat in having less lime and more soda. The real consanguinity of the rocks is obviously established. Taking into consideration the rocks of similar composition at Observatory Hill, Taylor's farm, and of other outcrops in the Fox river valley, to be described in a later bulletin by W. H. Hobbs and C. K. Leith, it will be seen that there is exhibited in these outlying pre-Cambrian igneous outcrops a petrographic province of considerable extent. The evidence of the probable Keweenawan age of the Baraboo rhyolite leads us to the conclusion that all the rocks belonging to this petrographic province are of the Keweenawan age.

Whether they were brought to the surface through a single vent or through many can only be conjectured. It is very probable, however, that the magma was transferred to the surface through a number of volcanic orifices and extrusive openings, and that the eruptive activity extended through a long period of time.

The isolated remnants of these great flows range in texture from volcanic breccia and surface flows to deep-seated granites, and also vary in amount of metamorphism that has taken place in them since they were brought to the surface. The problem undertaken in the working out of the geology of these isolated ledges is that of discovering the original condition of these congenital rocks and mode of occurrence, and also to trace the changes they have undergone since their solidification at the surface. In the present case the problem is pre-eminently one in metamorphism.

The mineral composition of these acid rocks is simple. Quartz and feldspar are the principal constituents, and since the former undergoes but little change, our attention is principally directed to the metamorphism of the feldspathic constituent and also to the change in the general texture of the rock.

The microscopical work necessary in the preparation of the

present paper has been carried on in the petrological laboratories of the University of Wisconsin and of the University of Chicago. I am under obligations to Professors C. R. Van Hise, W. H. Hobbs, and J. P. Iddings for kindly aid and criticism in the prosecution of this work.

I. THE UTLEY METARHYOLITE.

The outcrop of rhyolite at Utley is located in the central part of Sec. 36, T. 15 N., R. 13 E., in the southeastern part of the town of Green Lake in Green Lake county. (See map, Pl. I.) It consists of a single rounded knob-like area rising to an elevation of something over one hundred feet above the surrounding small valley of the Grand river, and is a noticeable and picturesque object in the small valley in which it occurs. The knob is about one-fourth of a mile in length, with a breadth of one-eighth of a mile. (Pl. III.) Its almost vertical slopes are rounded and worn into gentle contours by the polishing action of the ice of the glacial periods. Of late years the outcrop has supplied large quantities of macadam, quarries have penetrated it in various places, and it is a question of only a few years when the bluff will be planed down to the level of the surrounding valley.

The previous work done upon this area has been of a most general nature. Chamberlin,[1] in 1877, visited the outcrop, which was at that time known as the Pine Bluff porphyry, and described the topography and general appearance of the rock. A detailed study of the rock, of course, was not made by the earlier geological survey.

GEOLOGY.

Associated sedimentaries. The formations with which the rhyolite is associated are the horizontal deposits of sandstone and

---

[1] The Quartz-Porphyry of Pine Bluff, by T. C. Chamberlin: Geol. of Wis., Vol. II,, p. 250.

METARHYOLITE.   ST. PETERS SANDSTONE.   TRENTON LIMESTONE.   ALLUVIUM AND DRIFT.

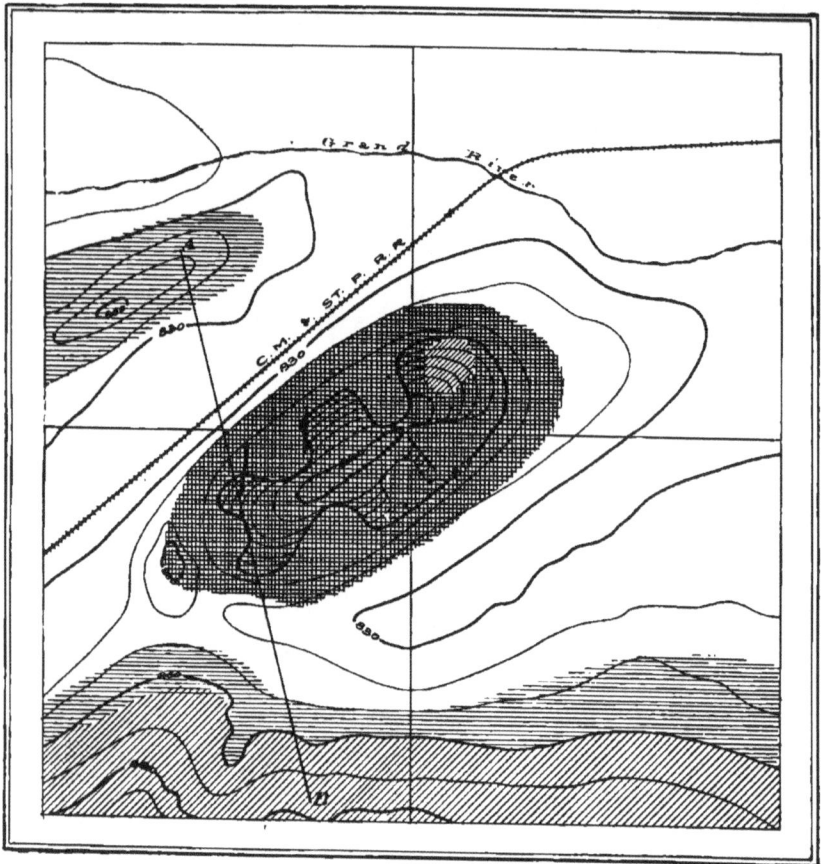

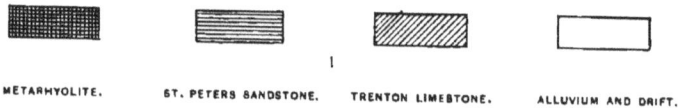

1. Geological map of the Utley metarhyolite area. The area mapped is ¹₄ sq. mile.

2. Section through the Utley metarhyolite area. Horizontal distance ¹₂ mile. Vertical distance through metarhyolite 100 feet.

limestone of the Ordovician system. Limestone, which presumably belongs to the Trenton formation, is in contact with the metarhyolite on its eastern slope near the summit. A short distance across the valley below this limestone the Saint Peter's sandstone outcrops in a low ridge. Both these sedimentary formations are horizontal and lie unconformably above the rhyolite. The relations of these formations are indicated in the cross section, Pl. III. The porphyry knob was thus an island in the Paleozoic seas, around which the sedimentaries were deposited.

Type of rock. In its composition and texture the Utley rock is a rhyolite, and since it is considerably metamorphosed it will be referred to as a metarhyolite, in conformity with the usage of the U. S. Geological Survey. The metarhyolite is massive, and of very uniform character. It is very dense and hard and easily breaks with a conchoidal fracture. The hard and brittle character of the rock is a quality of important economic value, for this rock, as stated above, is used extensively for macadamizing purposes. In the adaptation of the rhyolite for use the rock is crushed and broken into small fragments of various sizes, and the quality of glassy brittleness adds greatly to the ease of such breaking. In color it is uncommonly dark, the groundmass being almost black, in which are imbedded abundant phenocrysts of quartz and feldspar. The large and numerous crystals of pinkish white feldspar stand out in marked contrast with the black groundmass and give the rock a very pleasing and attractive appearance. The feldspar phenocrysts vary in size from a fraction of a millimeter to eight and even ten millimeters in diameter.

The quartz phenocrysts are about as abundant as those of feldspar, possess the usual limpid appearance, and most of them have good crystal form. The groundmass is very dense and appears almost glassy in the hand specimen, and usually does not reveal any flowage texture. The rock cannot be cleaved or split with ease in different planes and no evidence of secondary cleavage produced by pressure is visible megascopically. In places, however, fractures are quite abundant that are filled with secondary material, but such deformation is only local and has not affected the general texture of the rock.

Structure of the outcrop.   The layers of spheroids.   The outcrop has a structure due to the arrangement of the lava flows of successive extrusions.   Such arrangement of the successive flows is not to be detected in the normal rock, but is brought out distinctly by two layers of spheroids that are apparent on the northwest side of the knob.

The nature of the layer of spheroids is that of a thick mass of spheres or ellipsoids of rhyolite in a matrix of similar rhyolite. The character of the material composing the spheroids is exactly like that of the mass in which they occur.   The appearance of such a structure is shown in Pl. VI., fig. 1.   The origin of the spheroids lies in the common spheroidal parting, along which secondary alteration has proceeded in such a manner as to make the spheroids stand out as small nodules that are usually from an inch to an inch and a half in diameter.   In planes cutting through the spheroids alteration along the parting gives the appearance of circular zones of varying width which surround an inner core of rhyolite in appearance like that without the zone. The microscopic appearance of the spheroids will be dwelt upon later, and the relation of the spheroids to the perlitic parting will also be considered.

The upper layer of spheroids is well outlined by a thickness of about eighteen inches and could be easily traced from the base to the summit of the knob.   About twenty-five feet below the upper bed is another similar bed which could be traced for only a short distance on account of the quarrying of the rock and the accumulation of debris.   This lower layer lies in a plane parallel to the upper one and is much thicker than the one above, but is not composed of such densely packed spheroids.   Between the two layers, a distance of twenty-five or thirty feet, occur numerous scattered spheroids mingled with a large number of angular fragments of porphyry arranged with their longer axes parallel to the plane of the layers.   These fragments give the porphyry of this place the appearance of a volcanic breccia.   However, it is dense and massive and nowhere shows any trace of vesicular texture. The presence of the fragments in close association with the layers of spheroids indicates the surface conditions for the development of spheroidal parting. Below the lower layer, at the northeast end of the outcrop as well as above the upper bed at the southwest end, the rhyolite contains no spheroids nor fragments, but is of the usual normal type.

If the early history of the rhyolite formation is not misinterpreted, we have presented in this small outcrop the evidence that a cessation occurred in the general extrusion of the magma during the interval marked by the development of spheroidal parting in the upper part of the extrusion which forms the lower bed. This was followed by thin sheets of lava containing angular fragments of a similar rock with a sporadic development of spheroidal parting, accumulating to a thickness of about thirty feet. This volcanic breccia extrusion was followed by a period of quiesence in which the upper layer of spheroids was developed. This again was followed by another large extrusion of magma producing the normal rock of the north half of the outcrop. The beds of spheroids and volcanic breccia constitute not more than 5 per cent. of the whole outcrop, and dip about 60° to the southeast, the strike being somewhere near an east-west direction.

The absence of any marked deformation such as mashing has already been noted in the normal rock. It is also true that these beds of spheroids have been folded but little, if any, since their extrusion, as evidenced by the fact that they extend straight across the whole knob. The absence of appreciable dynamic metamorphism in the Utley formation will be referred to again in the comparison of the metamorphism of the three areas.

Intrusion of dikes. Two small dikes of greenstone intrude into the rhyolite at its northern part. The dikes run in parallel directions, striking 15° to 20° east of north, and have a dip of 85° to the west, cutting across the beds of spheroids. The larger dike has a thickness of two or three feet, and the smaller less than a foot. The greenstone is very fine grained, and is made up of plagioclase, amphibole, calcite, biotite, and chlorite. The plagioclase is very much clouded with decomposition products, and the minute crystals of hornblende, biotite, calcite, and chlorite are all very likely a secondary development after some ferromagnesian mineral, for their relations seem to point to a common origin. The mineral composition of the greenstone indicates that it is probably an altered phase of diorite or diabase.

MICROSCOPICAL PETROGRAPHY.

The thin sections of the metarhyolite show the groundmass to be lithoidal in character, very fine grained, but nowhere is it isotropic or glassy. The fine grained groundmass consists of feldspar and quartz and also several less important minerals of a secondary nature. Scattered throughout this aphanitic base are numerous phenocrysts of feldspar and quartz.

Textures of the rock. The metarhyolite presents several textures common to the volcanic rocks. Some of these textures are everywhere present, while others have but local development. In the discussion of the microscopic character of the rhyolite it becomes of importance to determine, so far as it is possible, the origin of the various textures—that is, to discriminate those phenomena which are of primary origin from those that are secondary. In such determination of the textures, the problem is extremely difficult. The processes that are effective in the crystallizing of rock magmas during their original solidification, are very probably not very different from those that develop the secondary phenomena. For example, the process of crystallization in producing an original lithoidal texture, is very probably similar, and the resultant texture is apparently the same as when devitrification is present, produced during a long period of time.

As seen under the microscope, the textures of the porphyritic metarhyolite are those of perlitic parting, spherulitic texture, poecilitic texture, and fluxion texture. Some of these have but a local development, while others, especially the fluxion texture, are present throughout.

Perlitic parting. Spheroidal parting, and the nature of the spheroids or nodular-like areas, has already been described in the hand specimen, and its occurrence in beds of the outcrop. There remains to be described the corresponding more minute structure known as perlitic parting as it appears in the thin section. An example of this structure is presented by the photograph of Pl. VI., fig. 2, in ordinary light. In the centre of the field is shown a circular line or zone about an inch in diameter, at one end of which is another smaller circle, in part bounded and wholly enclosed by the larger. Similar examples occur in other sections in which the perlites are smaller and more nearly spherical, and also some bounded by various curvilinear forms of different dimensions.

Perlitic parting, since it has its origin in rapid cooling and consequent contraction of an uncrystallized base, is usually considered as undoubted evidence of the once glassy character of the rock in which it occurs. The rock must also have attained a solid and rigid condition to allow the development of such a structure. But following these perlitic fractures, devitrification has been active, and a characteristic and persistent fibrous arrangement of the minute feldspar laths has been developed extending normal to the cracks. It has been noted in the perlitic parting of fresh glasses that minute fractures are induced normal to the major parting. Along these minute fractures these mineralizing agents could be introduced and re-crystallization would readily take place.

There are thus superimposed upon these original perlite cracks, zones of secondary fibres of feldspar and perhaps some of quartz, which are shown fairly well in Pl. VI., fig. 2. In this figure a dark line made up of minute specks of iron oxide and biotite marks the position of the original fracture. Sometimes in the smaller perlites these fibres of feldspar have so developed that they meet in the centre, forming spherulites whose origin has been from the outer boundary inward.

Mention has already been made of the spheroids as seen in the hand specimen and shown in Pl. VI., fig. 1, and the corresponding more minute structure of the perlites. It is believed that the spheroids, like the perlites, are due to spheroidal parting, caused by contraction during the process of cooling and crystallizing of the magma. The spheroids, however, show no radial arrangement of feldspar along the spheroidal parting, so far as was observed, but instead show a zone of weathering or alteration about one-fourth inch in width, which forms along the spheroidal parting. The weathered zone along the fractures is of lighter colored material than the general mass of the rock, as shown in the figure. The spheroidal fracturing allows the easy access of water solutions, and hence the alteration along these partings. The microscopic texture of the spheroids is like the general porphyritic texture of the rhyolite.

Well developed spherulites occur in the metarhyolite of Marcellon and Marquette. These, however, are not to be confused with the spheroids of the Utley area. Although there is a fibrous arrangement of the feldspar in the related finer texture of the perlites, yet in the larger spheroids there is an absence of radial arrangement of its constituents, while the spherulites of all sizes are made up of densely packed divergent fibres. In the weathered hand specimens the nodules and spherulites often look much alike, but on the fresh surface a difference can be noted, and with the thin sections under the microscope there is no need of confusing the two phenomena.

It very often happens that the fracturing of the rock, due to the contraction of the glass, does not produce complete circular forms, as seen in cross-section, but only part of circles and other curvilinear forms. In fact it is probable that many short curved fractures would be formed while but few perfect perlites would be developed. It is very often the case that triangular areas, as seen in section, are produced, bounded by three fractures that are concave inwards. Such concave triangular areas are as common, if not more so, than the circular partings of the perlites. Devitrification takes place along these short fractures and triangular areas, and there are developed curved strips and triangular areas of fibres as well as zones in circular shapes. Such a texture of devitrification products is shown to some extent in the above mentioned photograph, Pl. VI., fig. 2. In fact the texture here described is very similar to that noted by Mügge[1] in

---

[1] N. J. fur. Min., B. VIII., 1893, p. 648, Fig. 4, and Pl. 24 and 25.

the "Lenneporphyre" and termed by him "Aschenstructur" and which he explains as fragments of glass that have become devitrified or pseudomorphosed by feldspar and quartz. G. O. Smith[1] and F. Bascom[2] also describe structures similar to that described by Mügge as occurring in devitrified acid volcanics of Maine and Pennsylvania. Mügge,[3] who describes this structure quite fully, says of it: "Die Aschentheilchen sind keulenförmig und sickelförmig, auch ringförmige Stücke kommen vor. Ihre Füllsubstanz ist zum grossen Theil offenbar Quarz, zum anderen Theil Plagioklas. Dieser ist in manchen Aschentheilchen ringsum vom Rande her in Stängeln in das Innere hineingewachsen, während das Centrum von Quarzmosaik erfühlt ist (Tafel XXVII., Fig. 41)." The destription given by Mügge conforms closely, if not exactly, to the appearance of many of the sections of the Utley rock.

Rutley[4] refers to a similar structure in devitrified rhyolites, in a rock in which perlitic parting also occurs. In those rocks described by Smith[5] and also those by Bascom[6] perlitic parting is noted. While it is very possible that the explanation that Mügge offers is true for the "Lenneporphyre," yet in the Utley metarhyolite there seems to be a genetic relation between perlitic parting and the sickle-shaped triangular-concave and circular bands of feldspar lamellae.

Spherulitic texture. The spherulites are not numerous nor very well defined. They are of two kinds, those that have a well defined outer boundary, and those whose outer boundary is obscure. The first variety has been mentioned in connection with the perlitic parting. Where the zone of well defined fibres of feldspar occur along the perlitic cracks, and where these fibres are developed far enough to reach the centre, a typical spheru-

---

[1] Geology of the Fox Islands, Maine, by G. O. Smith: p. 39, Pl. I., Fig. 4.

[2] The Ancient Volcanic Rocks of South Mountain, Penn., by Florence Bascom: Bull. U. S. G. S., No. 136, p. 54.

[3] N. J. fur. Min. B. VIII., 1893, p. 713.

[4] Q. J. G. S., Vol. 37, p. 406, Figs. 1 and 2.

[5] l. c. p. 51.

[6] l. c. p. 55.

lite is produced. They have a distinct radial arrangement of the feldspar laths, but do not show the cross under the nicols as do some spherulites of original crystallization.

The second variety of spherulites is much larger than those just described. The fibres of feldspar, with perhaps some of quartz, radiate from the centre and do not have a well outlined boundary, but grade into the fine grained base. These are probably a product of the original crystallization, although they may be due to a process of devitrification.

Poccilitic texture. This texture is well shown in many of the sections. This mottled or patchy appearance of the groundmass common to many volcanic rocks is due to an intricate intergrowth of feldspar of slightly different orientation or of feldspar and quartz. It is usually considered of secondary origin, and the large amount of devitrification and metamorphism manifested in the Utley rock would seem to indicate that this texture is here very probably secondary.

Fluxion texture. The delicate and sinuous lines of flow so common in lavas, are everywhere present in this metarhyolite. In some sections it is well marked, and in others it is unimportant and ill defined. This texture is not apparent megascopically, as previously noted. It consists of streaks and lines of minute grains of iron oxide, biotite, and opaque specks. While there can be no doubt that the most of this texture, as it now appears, was produced previous to the original solidification of the rock, still there is also a great deal of it that is due to devitrification. This secondary process has developed and accentuated flow lines in places, and modified them more or less. The series of changes that has been referred to in describing the perlitic and so-called "Aschenstructur" undoubtedly tends to distribute the opaque material and iron oxide and secondary mica in irregular lines, either along perlitic cracks, or about older constituents of the groundmass. The minute flakes of biotite that have gathered about the corners and angles of the phenocrysts of feldspar, arranged in a parallel direction, give a general appearance of flow about such phenocrysts. The so-called "glass sherds" now devitrified, usually have an irregular parallel arrangement that tends to simulate an original fluidal texture. Of course, there can be no doubt that this rhyolite possesses a more or less original fluxion texture, but it is also very clear that a great deal of this texture is due to the parallel orientation of the secondary biotite about the phenocrysts and secondary devitrification along partings and fractures in the groundmass.

The feldspar phenocrysts. The nature of the feldspar phenocrysts is somewhat peculiar and of especial interest. Porphyritic minerals as a rule possess good crystal forms, but in the Utley rock most of the phenocrysts of feldspar have a wavy and rounded outline, and have apparent embayments and rounded extensions which project out into the groundmass. While the irregularity in outline of the phenocrysts may be due in part, as it often is, to magmatic corrosion, yet in this case it finds its explanation mainly in another source, which will be dwelt upon later.

Polysynthetic twinning, which is the common characteristic of plagioclase, is conspicuously absent in most of the feldspar of this rock. On the other hand, under a low power of the microscope, the phenocrysts appear to be mottled and have a patchy extinction. With the application of a higher power, this mottled appearance is seen to be due to an intergrowth of different varieties of feldspar, each part having a characteristic orientation.

As a rule the phenocryst is made up of but two varieties, but sometimes it has the appearance of being composed of three different species. These interlocking species of feldspar produce a typical perthitic structure which resembles that described quite fully by Brögger,[1] and also by Zirkel and Rosenbusch in their text books, under the terms perthite, microperthite, and cryptoperthite. Under these terms various forms of these intergrowths have been described, some quite similar to those occurring in the Utley rock, while others are quite different. Though the microperthite has the same general characters throughout the Utley rhyolite, yet it shows some variation, and there is a not-

[1] Zeit. f. Kryst. B. 16, pp. 521-560.

able difference between the Utley microperthite and that in the coarsely crystalline granite of Waushara.

Chemical composition of the feldspar. The composition of the feldspar is one of interest, and corresponds with the physical properties of the mineral, as well as the composition of the meta-rhyolite as a whole. Below is given the analysis[1] of the entire feldspar and also that of the rhyolite in which it occurs:

	I.	II.		III.
SiO_2	62.57	142.20	SiO_2	73.09
Al_2O_3	16.12	18.00	Al_2O_3	13.43
N_2O	7.20	11.61	Fe_2O_3	2.57
CaO	5.52	9.86	FeO	2.57
K_2O	5.11	5.43	CaO	2.28
Fe_2O_3	1.32	.80	MgO	1.03
			Na_2O	3.85
	99.84		K_2O	1.58
			H_2O	.72
				98.55

I. Feldspar.
II. Molecular ratio of constituents of feldspar.
III. Utley metarhyolite.

The material for the analysis was picked out by hand from crushed parts of the rhyolite that contained very large feldspar phenocrysts. These phenocrysts varied in size from one-eighth of an inch to one-half inch in diameter. They could be easily distinguished from the black groundmass and the crystals of quartz, for the latter constituent, like the feldspar, was composed of free individuals in the groundmass. A small amount of the groundmass, however, clung to the pieces of feldspar and was included in the substance analyzed. Less than 1 per cent. of quartz and 2 or 3 per cent. of groundmass is a large estimate of the foreign substance included in the above analysis. Of the lime-alkali constituent of the feldspar it will be seen that the soda is the most abundant, lime next, and potash the least. The

[1] The writer is indebted to Mr. C. F. Tolman of the Univ. of Chicago for the analysis of the feldspar.

molecular ratio of these three constituents is as follows: Na_2O : CaO : K_2O :: 11.61 : 9.86 : 5.43.

The feldspar thus is a soda-lime feldspar with a considerable amount of potash. We have seen that the feldspathic constituent is not a simple one, for it is made up, for the most part, of two species in close intergrowth. One species has the extinction angle and the composition of the soda-feldspar, albite, and the other species has the extinction of oligoclase-andesine. We have no reason to believe that any orthoclase is present, for the twinning habit and cleavage throughout is that of the triclinic feldspar. It seems very probable that the lime, soda, and potash enter into both species to some extent, although it is to be expected that the potash and most of the soda enters into the albite and the lime into the more basic species. The composition of the feldspar will be referred to again in discussing the origin of the microperthite.

A comparison of the feldspar analysis with that of the metarhyolite is given above. Since from 5 to 10 per cent. of the rock is made up of pure silica, as represented in the quartz phenocrysts, a fairly close agreement, as should be expected, is to be noted in the rhyolite magma and the feldspathic constituent.

Nature of the microperthite. A typical microperthitic feldspar of the Utley metarhyolite, as it appears in polarized light, is shown in Pl. VII., fig. 1. In this photograph the feldspar shows a good crystal form. The section is cut approximately in the plane of the brachy-pinacoid ($\infty P\breve{\infty}$) on the $\underline{M}$ face, and shows the cleavage parallel to the basal pinacoid (0 P). A bisectrix emerges almost normal in this section, and the angle of extinction for the adrk part was found to be from $+3°$ to $+6°$, and that of the light part about $+19°$. The dark part, although penetrating the parts not in total extinction in finger-like areas, is bounded by plane surfaces and has a greater length in the direction of the $\dot{c}$ axis of the phenocryst. The dark part, although not shown so well in the figure, is likewise bounded by prisms, domes, and pinacoids similar to the large phenocryst of which the two varieties are a part. The light part which extinguishes at an angle of $+19°$ has the extinction angle of albite of the

formula Ab.   The dark part with an extinction angle of from
+3° to +6° is oligoclase-andesine of the formula Ab_3An_1.

In fig. 1 there is shown a crystal of the microperthite twinned
according to the Mannebacher law, the composition plane being
parallel to the basal cleavage, and the section fortunately cut
parallel to the brachy-pinacoid ($\infty P \ddot{o}$) and almost normal to
a bisectrix.   In this crystal the relation of the perthitic intergrowth to the axes of symmetry is clearly indicated.   On the M
face the extinction angle for one part is +(17° to 19°) and

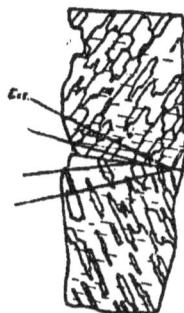

Fig. 1.—Crystal of microperthite, Mannebacher twins, cut parallel to $\infty P \bar{\check{o}}$, showing
the character and relation of the perthite inter-growths and their planes of extinc
tion.

for the other part +(5° to 7°).   On the $\underline{M}$ the extinction angle
for one variety is +(19° to 21°) and for the other +(5° to 8°).

The striations which appear in ordinary light, or the longer
axes of the perthitic growths, shown better under the crossed
nicols, make angles with the basal cleavage as follows:

On M 111° to 113°.

On $\underline{M}$ 119° to 121°.

This angle is the angle β, which in the acid plagioclase is
116° 25′ ±4′.   The apparent discrepancy is due to the fact that
the section is not cut exactly in the plane of the $\infty P \ddot{o}$ but is
slightly oblique to it.   To this obliquity is also due the slight
difference in the extinction angles of the perthitic growths of
the upper and lower twin.   That part which extinguishes at
+(17° to 19°) and +(19° to 21°) has an average of 19° which
corresponds with the plagioclase albite of the formula Ab.   The

part with lower angle of extinction $+(5°$ to $7°)$ is oligoclase Ab_3An_1. These small growths of albite and oligoclase, as a rule, have good crystal forms and interlock with one another in close intergrowth. They are bounded by prisms, domes, and pinacoids, and have a regular arrangement in the phenocryst which they compose.

A search was made through the thin sections to discover, if possible, whether there was any uniformity in composition of the perthitic growths as indicated by the angle of extinction. For this purpose phenocrysts were examined that were cut parallel to the brachy-pinacoid ($\infty P \ddot{\infty}$) showing basal cleavage and a normal bisectrix. At the same time it was soon noted that such sections always showed the perthitic growths with their longer axes parallel to the $\acute{c}$ axis of the phenocryst, which was another important aid in discovering the orientation of the crystal. The result of such examination of sections cut parallel to the brachy-pinacoid was as follows:

Microsection 4379.

One part has extinction angle of $+(19°$ to $22°)$, albite, Ab.

Other part has extinction angle of $-(3°$ to $8°)$, oligoclase-albite, Ab_2An_1.

Microsection 3850.

One part has extinction angle of $+(19°$ to $20°)$, albite, Ab.

Other part has extinction angle of $+(10°$ to $13°)$, oligoclase-albite, Ab_6An_1.

Microsection 3850 (another crystal).

One part has extinction angle of $+(17°$ to $19°)$, albite, Ab.

Other part has extinction angle of $+(4°$ to $5°)$, oligoclase, Ab_3An_1.

Microsection 3839.

One part has extinction angle of $+(19°$ to $23°)$, albite, Ab.

Other part has extinction angle of $+(6°$ to $11°)$, oligoclase, Ab_4An_1.

Microsection.

One part has extinction angle of $+(23°$ to $24°)$, albite, Ab.

Other part has extinction angle of $+(8°$ to $9°)$, oligoclase, Ab_3An_1.

Microsection 3847.   Fig. 1, Mannebacher twins.

Upper twin.   One part has extinction angle of $+(17°$ to $19°)$, albite, Ab.

Other part has extinction angle of $+(5°$ to $7°)$, oligoclase, Ab_3An_1.

Lower twin.   One part has extinction angle of $+(19°$ to $21°)$, albite, Ab.

Other part has extinction angle of $+(5°$ to $8°)$, oligoclase, Ab_3An_1.

From the above examination it can be seen that there is a close approach to uniformity in the angle of extinction of the perthitic varieties.   The apparent non-uniformity is due in part at least, to the section being cut somewhat oblique to the brachy-pinacoid.   But it is quite obvious that there are present as perthitic growths two well marked and distinct species of feldspar, one species being albite and the other oligoclase or oligoclase-andesine.   These intergrowths not only possess a typical crystal form but also have a common symmetry for both optical and crystallographic properties.   This common symmetry also, as indicated by the cleavage system, crystal outline, and twinning, is that of the phenocryst of which the perthitic growths are a part.

In sections that are not cut parallel to the brachy-pinacoid the growths appear to be irregular, due to the interpenetration of the two species and the merging of planes of extinction.   But for each crystal, however the section is cut, there is probably a regular arrangement of the growths.

The origin of the microperthite.   The origin of perthitic feldspar is a disputed question.   Some authors hold that such growths are original, a few that they are secondary, and others that the growths may be sometimes original and sometimes secondary.

Perthite was first described from Perth, Canada, as a new species of feldspar, but later it was found that this perthite was merely an intergrowth of orthoclase and albite and supposedly original.   Since this first mention of perthite numerous examples of its occurrence have been noted by many authors.   The perthites described in this paper, are of somewhat varying char-

acter and development, and present, it is believed, at least the
principal structures of perthite that have been described elsewhere. Since the conclusions arrived at are somewhat different
from those expressed by some observers, it is thought best to give
as full a description as possible of the occurrence in the Utley
metarhyolite, and the described adjacent areas, and then give a
resumé of some of the differing and concordant ideas held by
previous investigators.

To some extent the nature of the microperthite has already
been noted, but this will be further dwelt upon as the methods of
origin are described.

The development of the perthitic intergrowths has been subsequent to the solidification of the magma, and has been, in good
part at least, later than certain movements of the magma and
processes of alteration that have affected the rock. In the Utley
area this secondary development of the perthite has been brought
about in three ways, differing slightly from one another, yet
all by similar processes of secondary crystallization. These
processes have developed the intergrowths as follows: (1) As
secondary enlargements or outgrowths upon older crystals of
plagioclase; (2) As secondary intergrowths or regeneration of
older plagioclase; and (3) As complete phenocrysts in the
groundmass. These various developments were very probably
contemporaneous, whether built up from the groundmass or
from the already crystallized plagioclase.

(1) *As secondary enlargement or outgrowth.* In the photo-
graphs of Pl. VIII., figs. 1 and 2, are shown enlargements of
microperthite about older plagioclase. In fig. 1 the phenocryst
is shown under the cross nicols, and cut somewhere near the
plane of the macro-pinacoid, showing the albite twinning. The
figure shows an inner core of plagioclase of crystal form, surrounded by a thick margin of microperthite. The latter is
made up of small laths, wedges, and spindles of feldspar arranged with their longer axes in the direction of the c axis of the
inner core of plagioclase. This core, which shows the albite
twinning, is fractured and the fractures are filled with dark col-

ored material, biotite and iron oxide. The fractures and decomposition products of the core are shown better in ordinary light. The fractures are limited to the core crystal, and in minute pellucid areas crystals of calcite and sericite are scattered throughout the core, and there is an abundance of biotite in the fractures. These decomposition products and fractures are conspicuously absent in the rim of microperthite.

In Pl. VIII., fig. 2, is a similar crystal, in ordinary light, showing a zone of microperthite surrounding a core of plagioclase, the latter twinned according to both the Carlsbad and albite laws. The microperthitic rim is composed of laths with their longer axes parallel to the c axis of the core, and approach the orientation of the Carlsbad twins with which they come in contact. The inner core is seen to be filled with good sized black areas of biotite and iron oxide, and colorless crystals of calcite and sericite, and angular translucent dots. In the lower left hand part of the core, are several fractures running from right to left that do not penetrate into the outer rim. To the right, within the rim, is a large black area, iron pyrite, around which the microperthite has not wholly succeeded in growing. The dark areas within the core are wholly biotite.

Examples other than these were noted, showing enlargements of microperthite that did not wholly surround the core. Some few plagioclases of small size were found that were free from perthitic feldspar, and it is a noteworthy fact that without exception the plagioclase shows considerable decomposition, containing flakes of the micas and microlites, while the microperthite in the same section is of fresh appearance and free from alteration products. But in the above described sections, Pl. VIII., figs. 1 and 2, the true relation of the perthite to the elder constituents is clearly shown. The older phenocrysts are plagioclase and were fractured and considerably altered before the growth of the microperthite about them.

(2) *As secondary intergrowths or regeneration of older plagioclase.* This method of development is shown, to some extent in Pl. VIII., fig 1. In the lower left hand corner of the in-

ner core, above the dark line that separates the core from the rim, can be seen a mottled area like the perthite of the rim. This area extends upwards to the lower fracture of the core, and is made up of laths with their axes parallel to the $\dot{c}$ axis of the inner core, and is similar in appearance to the perthite in the rim.

In fig. 2 this method of change is also shown. In this figure there is presented a feldspar phenocryst of crystal form, without enlargement of perthite. The upper and left hand part of the phenocryst is microperthite without decomposition products. The lower and right hand part is twinned plagioclase conspicuously filled with alteration products, large crystals of brown biotite, and translucent microlites. The twinning of the plagioclase is according to the pericline law, the $\dot{c}$ axis running from right to

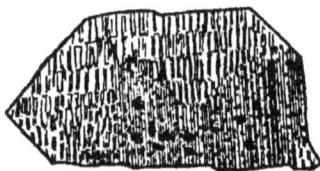

Fɪɢ. 2.—Feldspar phenocryst showing process of alteration from plagioclase to microperthite.

left in the figure. The pericline twinning may be secondary, further evidence of the secondary origin of which is presented in a similar phenocryst shown in fig. 3. In fig. 2 the section is cut somewhere near the plane of the macro-pinacoid, $\infty P \bar{\infty}$. The perthitic intergrowths appear as laths and spindles, and are arranged with their longer axes parallel to the pericline twins, and distributed in bands, from right to left in the figure, and seem to be governed by what may be the original albite twinning. It is impossible to tell from sections cut through the plane of the macro-pinacoid whether the orientation of the perthitic growths is the same as that of the older plagioclase or not. But it seems apparent that the arrangement of the perthite is governed by the twinning and symmetry of the older plagioclase.

In fig. 3 the relation of the original and secondary twinning and the microperthite is perhaps better shown. Figures 2 and

3 are of phenocrysts from the zone or beds of spheroids, which seems to have been an area of weakness in the rhyolite formation, for it shows more evidence of dynamic deformation, as well as weathering, than any other part of the outcrop.

In the crystal of fig. 3 there is evidence of deformation of a kind not seen in fig. 2, which is represented by the central zone of fracture marked d. The history of this phenocryst is thought to be about as follows. The original plagioclase crystal, twinned according to the albite law, was subjected to pressure, as evidenced by the zone of deformation and fracture across the middle, and probably produced the twinning at the ends of the crystal. Previous to the growth of microperthite, alteration to biotite, sericite, and microlites had taken place in the older twinned feldspar. The two areas marked a are the remaining

FIG. 3.—A bent feldspar phenocryst showing alteration of the plagioclase to microperthite.

traces of the albite twins, the two areas marked b are the remnants of the pericline twins, the parts marked c are microperthite, showing the laths of new feldspar which have developed in the planes of the pericline twins. The part marked d across the centre, shows fracturing, and, like the microperthite, is almost wholly free from secondary products. The zone of fractured feldspar, like the microperthite, owes its clearness to recrystallization, which process in the zone of fracture has been aided by pressure and mechanical movement, as is exhibited in the mashed clear feldspar phenocrysts of the Berlin rhyolite-gneiss.

We have thus seen in figs. 2 and 3, as well as in Pl. VIII., fig. 1, that the microperthite, as evidenced by its relation to the plagioclase, is a product of regeneration, as it were, of the original feldspar. This development is always perfectly regular and uniform, and is dependent upon the molecular arrangement of the pre-existing plagioclase. Other examples occur in the thin sections, very similar to fig. 2, that show clearly the genetic relation between the microperthite and plagioclase. In Pl. VIII., fig. 1, the laths of perthite are developed with their longer axes parallel to the $\overset{,}{c}$ axis of the older feldspar, and at the same time, apparently along the composition faces of the albite twins. In figs. 2 and 3 the perthitic growths are believed to be developed with their longer axes in the plane of the rhombic section following the composition faces of the pericline twins.

The microperthite represented by Pl. VII., fig. 1, and also the Mannebacher twins, fig. 1, already described, are examples of the regeneration of older plagioclase, although in these phenocrysts no trace of the original feldspar now remains. But the crystal form of these is such as to evidence the fact that the original feldspar was a simple plagioclase crystal. In these examples the perthite growths have their longer axes along the $\overset{,}{c}$ axis of the older crystal, each new growth being in parallel orientation with the original crystal. In the case of the pericline twinning, the new growths have their axes in the plane of the rhombic section, and it is very likely that they also are in parallel orientation with the original feldspar, although no definite plane through which the sections were cut, could be found so as to measure the optic orientation.

Some phenocrysts were noted that seemed to be made up of perthitic growth mingled with groundmass, the whole well bounded from the general groundmass by irregular boundary lines in shape like those shown in fig. 7. At other times irregular bands of perthite were noted cutting across phenocrysts which were probably older feldspar in its process of *perthitization*, although no cleavage, twinning, nor decomposition could be perceived in the part that might be the original feldspar. Sometimes the phenocrysts showed a rough zone or band, in which the perthite growths were irregular and not developed symmetrically with the principal part of the phenocryst. This is probably due to deformation at this place in the original crystal. Often in aggregates of feldspar crystals, as in fig. 4, the plagioclase, showing the usual twinning and containing numerous alteration products, stands out in marked contrast with the clear microperthite.

(3) *As complete secondary phenocrysts out of the groundmass.* The regeneration of older plagioclase into new microperthite has been described, as well as the development of the latter

Fɪɢ. 4.—Aggregate of feldspar crystals showing the clear appearance of the microperthite in contrast with the clouded plagioclase.

from the groundmass, as enlargement of older crystals. There now remains to be described growths of microperthite that have no genetic relation with pre-existing feldspar.

In fig. 5 is shown an irregular figure marked a, broad at the upper part, with a short projection above, and with a long tail-like appendage projecting downward. It consists of short feldspar laths with their longer axes normal to a dark line, a suture which widens in the broad part, enclosing a quartz mosaic. Below this and to the right, marked b, is another similar figure with four arm-like projections composed of feldspar laths normal to a median suture. This also has a broad part enclosing quartz, chalcedonic in appearance. Above and to the right, marked c, is a small phenocryst of microperthite, also possessing a median suture with laths of feldspar normal to it. The three central figures, a, b, and c, have a similar appearance and structure, and all extinguish together. The development of these areas is thought to be as follows: The dark line running through the bodies, the median suture, represents an early fracture in the groundmass, perhaps contractional parting due to original cooling, along which devitrification advanced, producing feldspar laths similar to zones of feldspar in the pearlite of Pl. VI., fig. 2, described on page 9. The crystal marked d̲, to the right, is a twinned plagioclase of original growth, showing the usual alteration products of the older feldspar. The large crystal of microperthite, marked e̲, is fresh and undecomposed, and, like the crystals in the centre of the figure, are in marked contrast with the alteration in the original plagioclase.

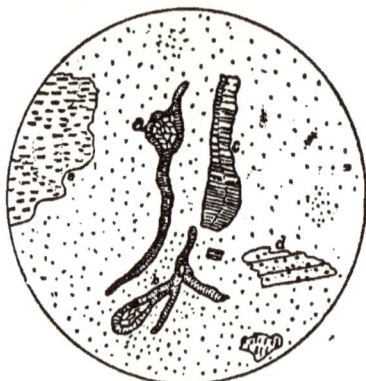

FIG. 5.—Section of the metarhyolite showing secondary growths of microperthite along fractures in the groundmass.

In fig. 6 similar growths of microperthite are shown. The growths marked a̲ and b are connected with a suture from which feldspar laths radiate normal to it, as in fig. 4. The relation of the dark line to the feldspar growth is well shown in the upper part of a̲. The central part of a̲ and b̲ is quartz. Above and to the left, marked c̲, is a similar growth having more the appearance of a phenocryst. Below and to the right, marked d̲, is a phenocryst of microperthite very similar to c̲, also with a dark line, normal to which laths of feldspar have grown.

It will be seen from figs. 4 and 5, that there is a close relationship between some of the small microperthite crystals and the irregularly bounded microperthite bodies with cores of quartz, which are probably built up along fracture lines, which latter are in turn closely related to the laths and spindles of feldspar along perlitic cracks. Since it has already been shown that the perthitic growths form secondary rims about the older plagioclase, and also secondary developments from the older feldspar itself,

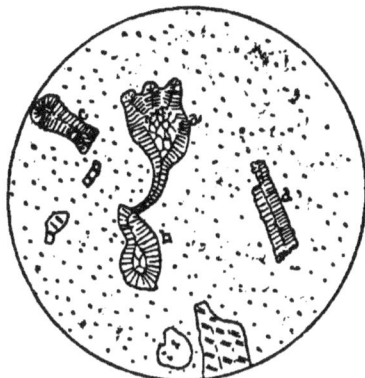

Fig. 6.—Section of metarhyolite showing secondary growths of microperthite along fractures in the groundmass.

it should be expected that complete new phenocrysts of microperthite would be developed from the groundmass. Many of the perthitic growths that possess irregular shapes, crenulated and wavy boundaries, without any traces of cleavage, or crystal faces, like those presented in fig. 7, are thought to be wholly of

Fig. 7.—Irregularly shaped phenocrysts of microperthite.

secondary growth. A distinction, of course, cannot always be made between those of entire new development from the groundmass, and those that have enlargements with the cores also *perthitized*, for the growth within the core is exactly similar to that without, and unless the dark material about the core remains, no traces of the older feldspar are left. But if the phenocryst is cut through a characteristic crystallographic plane so as to show cleavage and the planes of extinction, differences can be detected, for those of secondary growth show no cleavage system, and although there is a uniformity in the optic orientation of the new growths, there is no accurate way by which to measure the latter, so far as could be made out. As above stated, there is a regularity in the growth of the secondary phenocrysts, each having symmetrical growths of laths of feldspar with good crystal form. These laths are developed with their longer axes normal to the longer axis of the resulting phenocryst, in many cases at least, but this may not be a general rule.

FIG. 8.—Section of metarhyolite showing alteration of feldspar along cleavage planes.

It may be suggested that these small laths of slightly different extinction are only twins of the same species of feldspar, but their similarity of growth and appearance to the *perthitized* feldspar in which two species are invariably present should be evidence that all the mottled feldspar is made up of two species.

Weathering and fracturing of the feldspar. Weathering, as represented in fig. 8, along fractures and cleavages cracks of the feldspar is characteristic of the beds of spheroids. In the process of weathering the spaces within the microperthite, like those shown in fig. 8, are filled with sericite, biotite, calcite, and iron oxide. Usually the feldspar has not been broken apart, but these secondary minerals appear to be alteration products of the feldspar in place.

In several places of the outcrop are numerous fractures varying in size from a fraction of a millimeter to ten or twelve millimeters across. These fractures are filled with vein material of quartz and feldspar. Where the fractures have passed through feldspar, fig. 7, there the vein material is feldspar of similar appearance and orientation, while the fractures through the quartz are filled with quartz likewise of similar orientation. In some cases vein material one-half inch in thickness was wholly composed of feldspar where such veins were through phenocrysts of feldspar. The fractures through the groundmass are usually filled with quartz, although feldspar is also often present.

In the fractured crystals where the vein feldspar, as well as the vein quartz, have the same orientation as the penetrated crystal, there is shown an excellent example of the effectiveness of the energy of crystallization in the selection and orientation of material. When the crystal becomes fractured and mineral matter in solution fills the fissures and cracks, the broken crystal exerts a choice and controls the development of the vein material in that part. The feldspar or quartz is healed much like the healing of organic units. It is this energy of crystallization that is the controlling factor in the development of the perthite in the original plagioclase, so that the new feldspar species within and about the older feldspar have a uniform and regular arrangement dependent upon the symmetry of the parent crystal.

Quartz phenocrysts. The phenocrysts of quartz are less abundant than those of feldspar, and they possess a better crystal form in general than the latter. Besides the usual appearance of quartz, there is noted in these quartz phenocrysts certain outgrowths or enlargements of a secondary nature. These enlargements are not rims like those of the microperthite, but irregular fringe-like projections that extend from the phenocryst out into the groundmass. These growths are fairly well shown in fig. 9. In these examples the original crystal form of the quartz is well developed, and, projecting from them in irregular growths with identical optic orientation, is the secondary quartz. Sometimes a dark line marks the boundary of the original crystal. At other times the crystal planes must be prolonged to indicate the original boundary. Secondary quartz as vein material having a common orientation with the fractured quartz has been noted above.

FIG. 9.—Phenocrysts of quartz showing secondary enlargement.

The quartz phenocrysts often have inclusions of groundmass which are sometimes rounded areas, but very often they are rhombic in form. Rhombic cleavage of a fairly persistent nature can be noted in many of those quartzes that are ground quite thin. Like some of the feldspars the quartzes often show embayments, probably due to magmatic corrosion before the original solidification.

Amphibole phenocrysts. Although amphibole is not now present as a porphyritic mineral, yet that it once occurred as such is evidenced by characteristic areas now composed of alteration

10.   11.

FIG. 10.—Basal section of altered amphibole.
FIG. 11.—Prism section of altered amphibole.

products of the original hornblende. In fig. 10 is shown one of these areas which is without doubt a characteristic basal section of amphibole. Fig. 11 is an area composed of similar minerals which is thought to be a prism section of what was once an amphibole. The chemical composition of the rock, as indicated by

the presence of lime and magnesia is such as to allow the formation of amphibole from the magma. No trace of the original
hornblende remains except the form of the crystal, the space it
once occupied being now filled with a number of new minerals.
Six minerals were noted in the basal section represented in fig.
10. In other areas other minerals were noted, so that at least
seven minerals were detected assuming the space of the original
amphibole. Among the minerals noted were biotite, sericite,
quartz, magnetite, epidote, sphene, apatite, and zircon. The
boundary of the area is usually marked by a line of magnetite
crystals, while the quartz seems to fill the central part.

*The minerals of the groundmass and alteration products of
the phenocrysts.* The feldspar and quartz of the groundmass
which make up its greatest proportion have already been referred to in the description of the various textures of the rock.
The phenomena of the secondary enlargement of the feldspar
phenocrysts by perthitic growths, and the development of fibers
of feldspar along cracks in the rock would seem to indicate that
much of the feldspar of the groundmass is secondary. The
quartz often occurs in veins and as growths about older phenocrysts of quartz. Of the remaining small minerals scattered
throughout the crypto- crystalline groundmass, many are secondary, and many of them may be original. In many cases the relative positions of these minerals evidence their secondary nature;
in other cases no such criteria are at hand. The extreme metamorphism of the feldspar phenocrysts and the evidence of the devitrification of the groundmass are such as to indicate that the
common basic minerals so often secondary elsewhere have a
metamorphic origin here.

Biotite. Biotite occurs in considerable abundance throughout the groundmass, distributed irregularly throughout the latter, and also in streaks and along fracture lines. It also occurs
in considerable abundance as an alteration of the original plagioclase in the fractures and along the cleavage planes of the latter.
It is of very uniform habit, being in the form of short and thick
tabular crystals of quite uniform size. It often occurs in

streaks about the corners of phenocrysts, giving them the appearance of eyes. Biotite, next to quartz and feldspar, is the most abundant of the minerals of the groundmass.

Sericite and muscovite. These minerals, like the biotite, occur in both the groundmass and in the phenocrysts, and are more abundant where the weathering of the rock is more advanced.

Calcite. Calcite may be in part an infiltration product from the adjacent limestone formation. It, however, occurs more often in the feldspar phenocrysts than elsewhere, where it occurs in the central parts of the feldspar.

Hornblende. Brown hornblende, closely associated with the biotite, occurs to a small extent. It has much the appearance of the brown biotite, and can be detected only by its polarizing colors.

Sphene. Sphene occurs more abundantly in those sections that showed the traces of original amphibole phenocrysts than elsewhere. Here it occurs in quite large twinned crystals in the groundmass, and in smaller crystals associated with the alteration products of the hornblende.

Epidote. This mineral occurs to some extent as an alteration product of the amphibole phenocrysts.

Apatite. Minute needle-like crystals, colorless and with cross cleavage or parting, are probably apatite.

Zircon. Zircon with characteristic form and appearance occurs as an alteration of the amphibole and is also present to a small extent in the groundmass.

Iron oxides, magnetite and pyrite occur to some extent. Some of it may be original, but much at least is of secondary origin.

Quartz and feldspar as alteration products have already been dwelt upon.

II. THE BERLIN RHYOLITE-GNEISS.

As previously stated the Berlin area of gneiss consists of a single outcrop, and is located in the northern part of Green Lake county. It lies within the limits of the small city of Berlin and forms a rather prominent elevation, rising to a height of about 150 feet above the adjacent level valley of the Fox River.

The accompanying map, Pl. IV., shows the areal extent and topography of the Berlin gneiss and its relation to the surrounding rock formations. The area is located in the N. E. 1-4 of Sec. 3, T. 17 N., R. 13 E., in Green Lake county. The outcrop consists of three elongated domes connected in the form of a curved crescent-like ridge whose convex side faces the northwest and whose points are directed approximately to the south and to the east. In areal extent it is about one-half a mile long and it is about one-eighth of a mile wide.

The previous work done upon this area has been of the most general character and is embodied in the various reports of the early geological surveys.

Percival[1] in 1856 alludes in a general way to the topography of this area and gives a brief description of the rock. Chamberlin,[2] in 1877, in describing the Berlin Porphyry notes the general character of the rock in the field, its cleavage system, and extensive fissures, and also its relation to the adjacent overlying Potsdam sandstone. Irving[3] in 1877 speaks briefly of the Berlin rock. He notes the dark bluish grey matrix, numerous feldspar crystals and the lamination which is said to be fine and distinct.

GEOLOGY.

Associated sedimentaries. The only sedimentary formation, other than the glacial drift, with which this rock comes in contact, is the Potsdam sandstone and conglomerate. Lying upon

---

[1] Annual Report of Geol. Survey of Wis. for 1856, p. 106.
[2] Geol. of Wis., Vol. II., pp. 249-50.
[3] Geol. of Wis., Vol. II., p. 521.

RHYOLITE-GNEISS.   POTSDAM SANDSTONE.   ALLUVIUM AND DRIFT.

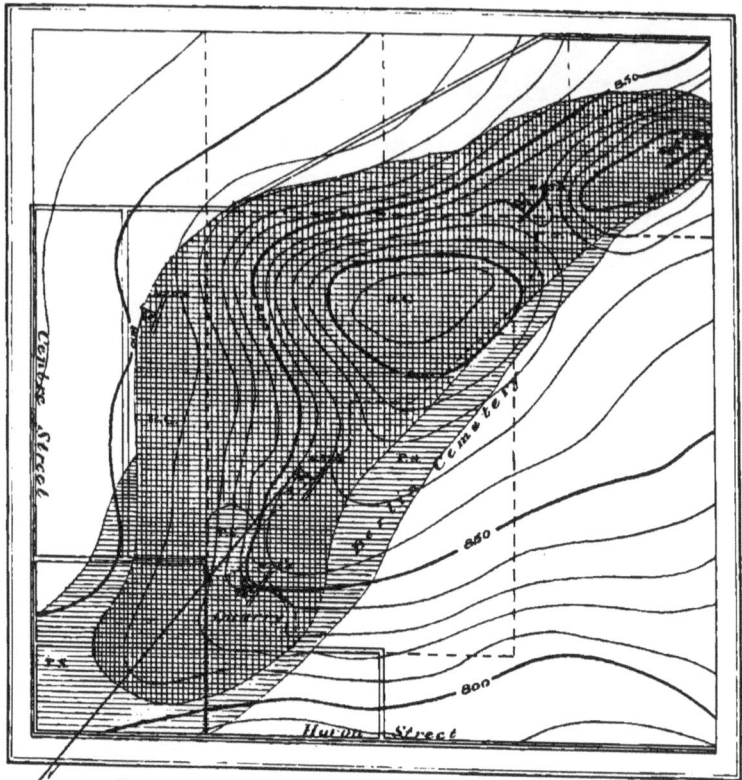

1. Geological map of the Berlin rhyolite-gneiss area. The area mapped is ¼ sq. mile.

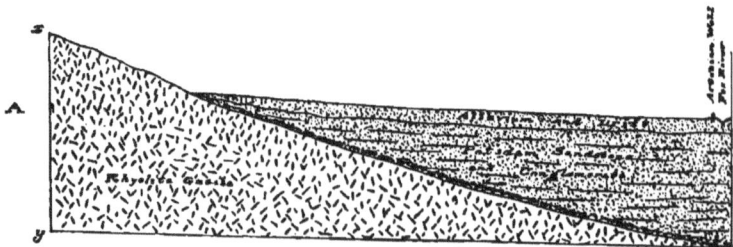

2. Section through Berlin area. Horizontal distance 1 mile. Vertical distance X to Y,
600 feet; N to M, 431 feet.

the southeast end of the gneiss is a small patch of conglomerate which is composed of angular and rounded fragments of the underlying igneous formation, mingled with a loose cement of sand. This conglomerate fortunately is fossiliferous and carries several species of brachiopoda and trilobita, among the latter being a species of Dikellocephalus. The fossils indicate the probability of this formation belonging to the Potsdam, although it is near the same elevation as the Magnesian limestone which occurs a short distance to the east. There is sandstone outcropping also along the southeast border of the gneiss formation.

A thick mantle of glacial drift and alluvium overlies the Potsdam formation about the base of the gneiss, and in places the drift also overlaps the igneous formation.

The thickness of the formations overlying the gneiss is indicated in the cross section of Pl. IV. These data were obtained on the sinking of an artesian well three-fourths of a mile southeast of the south end of the gneiss. The depth of this well is 456 feet, the first sixty feet being in glacial drift and alluvium, followed by 371 feet of Potsdam sandstone and conglomerate, the last twenty-five feet of the well being driven into the gneiss. The well is located on the bank of the Fox River near the railroad depot, which is 179 feet lower than the summit of the outcrop of gneiss. This would make the summit of the outcrop possess an elevation of 616 feet above that point in the well where the contact between gneiss and conglomerate is located, or in other words that this formation was an elevation of something over 600 feet at the time of the encroachment of the waters of the Potsdam sea.

Type of rock. The Berlin gneiss is a metamorphic rock, the evidence of which will be brought out in the study of the rock as seen in both the field and in the laboratory. The original rock was a rhyolite and now has the texture of a gneiss, and therefore it is referred to in the following as rhyolite-gneiss.

The gneiss possesses a marked uniformity of character throughout the area. The fresh surface of the rock reveals an almost black groundmass streaked with red. Thickly sprinkled throughout the groundmass are numerous white feldspar crys-

ROCKS 3.

tals of various sizes, the more prominent ones ranging from two to five millimeters in diameter. The phenocrysts of whitish feldspar stand out in marked contrast with the dark groundmass, producing a beautiful and striking appearance. They have their longer axes in a parallel direction, in the plane of the banding, and are often broken and drawn apart, as is shown in the lithograph of the specimen, Pl. V. Phenocrysts of quartz, so abundant in the Utley rock, do not occur in the Berlin gneiss. Although the groundmass is of very fine grain, the rock does not break with a conchoidal fracture because of the pronounced cleavage everywhere present.

Banding or lamination parallel to the cleavage is brought out distinctly on the weathered surface. The bands are dark and reddish streaks of fine groundmass, this dark color being due to microscopic crystals of hornblende disseminated in streaks throughout the rock.

The Berlin rock has been subjected to the profoundest kind of metamorphism. A perfect system of cleavage producing a pronounced banding or lamination has been induced in the rock by orogenic forces. Accompanying the movement that produced the cleavage there has been a marked change in the texture of the rock, as seen in both the hand specimen and the thin section.

The gneiss is believed to have been a rhyolite for the following reasons: Throughout the rock the idiomorphism of the porphyritic crystals of feldspar is a marked characteristic. Although these phenocrysts show variable amounts of deformation, yet they can generally be easily distinguished from the groundmass in which they occur. And in places in the gneiss where the mashing has not been complete and the lamination is indistinct, the texture of the rock is that of a rhyolite very like that of the Utley outcrop. These incompletely mashed patches in the gneiss, seen in cross section, usually have an area about the size of a man's hand, which areas have irregular elongated shapes. They grade into the more completely laminated gneiss, and are especially apparent in the plane normal to the longer axes of phenocrysts, the "head" of the quarried blocks.

BERLIN RHYOLITE-GNEISS.

NATURAL SIZE.

Whether the rhyolite-gneiss was originally a surface flow or an intrusive mass is not so clearly shown. It should be noted that the other outcrops of the rhyolite in the Fox River valley, as well as the one at Baraboo, appear to be surface flows, and most of them are associated with volcanic breccia. And in this connection may be mentioned certain angular fragments that occur through the Berlin area. These fragments vary in size from a few inches to almost a foot in diameter, and are of somewhat darker material than the general mass. Like the phenocrysts these fragments are flattened in the plane of the cleavage, although their parallel arrangement in this plane is not so complete as that of the phenocrysts. This incomplete orientation is especially true of some of the larger fragments. The fragments have been subjected to as much deformation as the normal rock, and evidence the fact that they were present in the rhyolite before the mashing process began. They are scattered promiscuously through the formation, and are not associated with any zone of faulting nor with any foreign rock. While the evidence of the extrusive origin of this rock is not conclusive, yet from its analogy with the other rhyolite outcrops, and from the evidence of the fragments, the Berlin rock is believed to be a surface rock like the other rhyolite outcrops of the region.

While deformation of the rhyolite is chiefly that of mashing, there was a concomitant development of a system of fractures and joints. The deformation is in the zone of combined flowage and fracture, as defined by Prof. Van Hise,[1] but principally in the zone of flowage, as will be shown in the petrographic description of the rock. Evidence in this line is also brought out in the field in the fact that nowhere is the rock fissile, for with the mashing and flowage process the rock was completely healed and possesses only the capacity to split in certain directions designated as the cleavage planes.

Structure of the outcrop. The outcrop can be said to have a definite structure which is the pronounced banding or cleavage system of the rock and its associated fracture planes and joints.

[1] Principles of North American pre-Cambrian Geology, by C. R. Van Hise, 16th Ann. Rept., U. S. Geol. Survey, Part I., 1896, pp. 601-603.

The cleavage system.   The strike and dip of the cleavage planes, as indicated upon the map (Pl. IV.), change gradually in passing from the southwest end of the crescent-shaped area to the northeast end.   At the southwest end the strike of the cleavage planes is N. 37° E., and the dip is 70° N. W.; in the central portion of the area the strike is N. 45° E. and the dip is 68° N. W.; while at the northeastern part of the ridge the strike is N. 70° E., and the dip is 65° and 68° N. W.   The strike of the cleavage is always further to the east in passing from the southern end to the eastern end of the crescent-like ridge.   The strike of the cleavage thus coincides with the axis of the ridge, the crescent shape being due to weathering and erosion processes which have been controlled by the system of secondary cleavage.   Farther away from the main axis of the ridge, to the northwest, near Greenwood St., the strike is N. 20° E. and the dip is 80° N. W. Numerous measurements were made at various points on the ridge showing the dip to vary from 65° to 80° with no abrupt but always a gradual transition between various places.

Joints and fracture planes.   The joints and fractures belong to more than one period of deformation, as evidenced by the fact that many of them are completely cemented, while others are not.   Those joints that are wholly cemented belong to an earlier period and appear to bear some relation to the cleavage system of the rock.

The fractures of the earlier period in a general way strike in four directions, and these four directions appear to be closely dependent upon the strike of the cleavage system for they change in direction as the strike of the cleavage changes in passing from one end of the crescent-like ridge to the other.   The dip of these cemented joints is much more variable than the strike and no definite general dip could be made out although they usually have an inclination of about 70° to 80° on either side of the vertical plane.

The strike of some of the fractures coincides with the plane of the cleavage.   The larger number of fractures, however, are approximately at right angles to one another and strike about 45°

to the plane of cleavage. There are also numerous small fractures that strike normal to the cleavage plane.

Numerous excellent examples showing the relation of the planes of fractures and plane of cleavage were noted in the area, and measurements made, with the results in general as above stated. It was attempted to photograph the best examples to illustrate this phenomena, but the attempt unfortunately was not successful. A sketch, however, was made at one place which is here presented. (Fig. 12.)

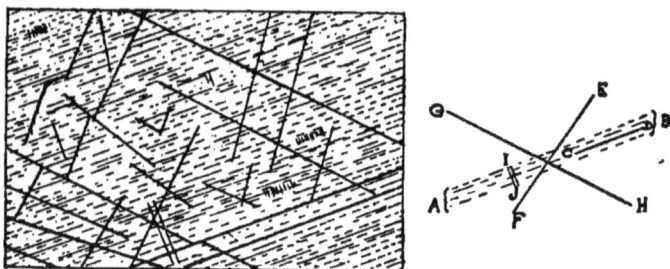

FIG. 12.—Sketch of horizontal surface of rhyolite-gneiss showing relation of earlier system of joints to the cleavage system.

The sketch is of the horizontal surface of a ledge near the northeast end of the area. The joints run in four directions. The plane of cleavage (A. B.) at this place strikes N. 75° E., and one set of joints (C. D.) coincides with the cleavage. One of the diagonal sets (E. F.) strikes N. 40° E., thus cutting the cleavage at an angle of 35°. The other diagonal set (G. H.) runs N. 60° W., and cuts the cleavage plane at 45°. Another set (I. J.) made up of shorter fractures cuts the cleavage plane at 90°. There are also present numerous random fractures that do not have any definite general direction and are not at all persistent like those above enumerated, but are wholly cemented like the above.

Zones of numerous closely packed fractures often occur, which mark the places where movement of the rock was concentrated. These zones usually cut the cleavage plane at angles varying from 30° to 50°. It is probable that the cemented fractures were produced during the latter part of the period of deformation that developed the cleavage in the rock. This view appears to be evidenced by the fact that the most numerous joints and zones of fractures strike diagonally across the cleavage, as represented by E. F. and G. H. of fig. 12.

As a result of experiment it has been clearly demonstrated[1] that a solid, brittle body when subjected to compression greater than its elastic limit, tends to fracture along planes of maximum shearing, which planes in a homogeneous body approach an angle of 45° to the pressure, and about 90° to one another. Since most authors hold[2] that cleavage develops in a plane normal to the greatest pressure, it would appear that the fractures that strike diagonal to the cleavage are probably in the planes of maximum shear. The fractures parallel to the cleavage are likewise probably due to shearing and are directed into the plane of cleavage because the latter is a plane of weakness in the rock, for the parallel fractures, if followed for some distance, are seen to branch off and cut across the cleavage at about 45°. The short discontinuous fractures transverse to the cleavage appear to have been produced at the same time as the diagonal joints. These transverse fractures are partly controlled by the cleavage of the feldspar (see p. 41) and are probably the result of tensile stresses resulting from the compressive stresses parallel to them, which produced the major diagonal joints already mentioned.

In addition to these earlier and healed fractures there are numerous open joints that appear to have no relation to the cleavage, but cut one another and the cleavage at all angles. They cut the rock into large irregular blocks, and are of a much later origin than the healed joints associated with the cleavage.

MICROSCOPICAL PETROGRAPHY.

The microscopical study of the Berlin rock, like the field study, reveals a marked uniformity of texture and grain of the gneiss throughout the area. As previously stated the gneiss is banded or laminated, consisting of a fine-grained dark ground-

---

[1] Geologie Experimentale, A. Daubree: Vol. I., pp. 391–432.

[2] Principles of North American pre-Cambrian Geology, by C. R. Van Hise: 16th Ann. Rept., U. S. Geol. Sur., Part I., 1896, pp. 633–668.

mass inclosing numerous phenocrysts of whitish feldspar. The groundmass has but one characteristic texture and that is a fluxion texture due to the mashing of the rock by extreme pressure which took place after its original solidification.

The evidence of the mashing of the rock is shown in the field by the presence of the persistent cleavage and fracture system throughout the area. Evidence of mashing is likewise borne out by the study of the rock under the microscope, as will be shown. The individual constituents of the groundmass, however, the quartz, feldspar, and minute flakes of hornblende, in themselves, reveal no evidence of deformation although mashing is indicated in their arrangement in the banded structure. The phenocrysts on the other hand show very clear and abundant evidence of deformation. For this reason the porphyritic constituent of the gneiss will be described first and will be followed by the investigation of the character of the groundmass.

The feldspar phenocrysts. The only porphyritic constituent of the gneiss is feldspar. Quartz as phenocrysts does not occur. The feldspar crystals are numerous and vary in size from 1 mm. to 5 mm. in diameter. A crystal of the feldspar was detached from the rock, and the angle which the two most important cleavage faces make with one another was measured. The angle was measured by means of the reflection goniometer, and was found to be $86° 28\frac{1}{3}'$.

The specific gravity determination was not satisfactory. The Thoulet solution at 2.60 allowed about one-half of the feldspar to float and the remainder to sink and be suspended in the liquid. This variation was probably due to inclusions of hornblende or magnetite within the feldspar.

No chemical analysis of the Berlin feldspar was made, but since the composition of the rock closely agrees with that of the Utley rhyolite in which the feldspar proved to be albite-oligoclase, it is most probable that the Berlin feldspar is likewise a plagioclase near the albite end of the series. The conclusion is also borne out by the cleavage angle $86° 28\frac{1}{3}'$ between 0P and $\infty P\check{\infty}$ which is that of acid plagioclase. Many of the phenocrysts of feldspar have been so deformed by the mashing process, in being fractured and granulated, that few determinations of the optical proporties could be made. However, those that do not show the effects of the pressure, manifest clearly that habit of microperthitic intergrowths characteristic of the Utley metarhyolite.

On one of the crystals that consisted of Mannebacher twins the angle of extinction was measured against the basal cleavage, and showed the two perthitic intergrowths to be essentially the same as those represented in the Utley microperthite. One variety extinguished at +4° to +5°, and the other at +19°, indicating oligoclase and albite respectively. The feldspar usually appears to be fresh, that is, it is clear and does not as a rule contain any of those microlitic mineral inclusions and alteration products so common in the process of weathering.

In Pl. IX., fig. 2, is shown a feldspar about 1-12 of an inch in length. It manifests very clearly the effects of pressure in the fractures and irregular extinction that characterizes it. The fractures are along the shortest diameter of the crystal, and are of various sizes ranging from the several plainly visible in the photograph, to minute lines of microscopic and sub-microscopic size. Near the lower end of the crystal, on the right side, granulation has begun. The slightly irregular boundary of the crystal also manifests this phenomena of the breaking down and recrystallizing of the parts of the phenocryst. The most interesting phenomena of this phenocryst, however, are its fractures and strain shadows.

In Pl. IX., fig. 1, are presented two crystals, the one above having been pressed down upon the one below and broken it apart. The upper crystal is fractured and granulated near the middle part. The principal characteristic of this crystal, however, is its fine cross hatching which resembles in a remarkable degree that of typical microline. Immediately below the centre of the crystal is an area of well-defined oblique hatching which seems to be superinduced upon the normal cross hatching. The normal hatching is supposedly due to the cross twinning, according to the albite and pericline laws,· but the oblique hatching, equally as symmetrical, cannot be ascribed to either of these.

Parts of the crystal show no twinning at all, while other parts, as noted, show both normal and oblique twinning. These differences without doubt are due to unlike strains induced in the crystal rather than any difference in composition. A slender area of granulated feldspar and coarse groundmass projects from the left end of the crystal. The underlying broken crystal has a filling of quartz principally between the broken parts. This quartz has a wavy extinction, and indicates that stress was here applied after the crystals of quartz had formed between the sundered parts of the phenocryst. Between the two crystals is granulated feldspar and blue amphibole. At either end of the crystal the effects of the granulation and recrystallization are seen in the short tail-like areas that project from them in the plane of cleavage.

In Pl. IX., fig. 4, is shown a part of a phenocryst of feldspar that has been drawn out to several times its original length. The crystal shows fractures and strain shadows. The spaces between the several parts are filled with quartz, plagioclase, and amphibole. The transverse fracturing is probably controlled by the cleavage of the feldspar and is probably due to tensile stresses resulting from compressive stresses normal to the longer axes of the phenocrysts. Numerous short fractures transverse to the cleavage were noted in the field that probably have an origin in common with the lengthened phenocrysts. Other phenocrysts show different stages of granulation, as shown in Pl. IX., figure 3, while some of them are apparently entirely crushed, leaving no traces except areas of quartz and feldspar considerably coarser than the average groundmass. Others show but little deformation except the usual short or long tail-like area that projects from the phenocryst in the plane of the cleavage.

The groundmass. The groundmass consists of granular quartz and feldspar in about equal proportions, and numerous small crystals of blue amphibole. The amphibole is a minor constituent, but it has a uniform distribution. Besides these three important minerals there is present in subordinate quantity calcite, biotite, garnet, zircon, apatite, and magnetite.

The quartz and feldspar. About 95 per cent. of the groundmass is made up of quartz and feldspar, which occur as small,. closely fitting grains. Both minerals are fresh and clear, and for this reason it is difficult to distinguish them except by the twinning habit of the feldspar. The feldspar is a variety of plagioclase, as indicated by its polysynthetic twinning habit, and very often it has the cross twinning common to microcline and is probably soda microcline. The size of the feldspar and quartz as compared with the phenocrysts, and their general habit and arrangement, are brought out in Pl. VII., fig. 2, and Pl. IX.

The minute grains of quartz and feldspar of the groundmass. stand out in sharp contrast with the larger phenocrysts. They vary somewhat in size, and seen in cross-section their diameters average about 1-40 mm. in length. There is a slight difference in the length of the diameters of these small crystals, and it is apparent on a close measurement that they have their longer axes in a common direction. In the process of the granulation of the phenocrysts quartz is mingled with the granulated feldspar, and in the spaces between the stretched phenocrysts the filling consists of quartz and feldspar in about equal abundance.

The blue amphibole. The amphibole occurs in very small crystals arranged with their longer axes in a common direction. It is uniformly distributed throughout the area, and it is to these minute crystals and the very small amount of iron oxide present that the dark color of the gneiss is due. It occurs in well-detined crystals, the prismatic planes ∞P and $\infty P\ddot{o}$ being usually present. The prism faces are well defined and the terminal faces are also well developed in the very minute crystals, but the ends of the larger crystals as a rule are jagged and frayed out. The perfect cleavage parallel to the prism faces is well shown.

Repeated efforts were made to determine the position and intensity of the axes of elasticity, but without success. The small size of the crystals and the strong pleochroism is such that nothing definite could be ascertained. The amphibole is deeply colored and strongly pleochroic, the difference in color absorption in the direction of the three axes being always very prominent. The absorption colors along the crystallographic axes à, b̄, and ĉ are as follows: along ĉ = deep blue, along b̄ = azure blue, along ā = light yellow green. This would assign the amphibole to any one of the three blue varieties, glaucophane, arfedsonite, or riebeckite. These are the amphiboles especially rich in soda, and would be expected in a rock like the Berlin gneiss which contains a considerable amount of this alkali. Riebeckite differs from glaucophane in having c nearly coinciding with crystallographic à, and a nearly coinciding with crystallographic ĉ, while in glaucophane the axes of elasticity are nearly the corresponding crystallographic axes. Since the relative position of the axes of elasticity could not be determined nor an analysis made, it was impossible to identify the mineral closer than to place it as a member of the blue soda-amphibole series.

The blue amphibole occurs in streaks and aggregates in straight lines, and where adjacent to phenocrysts they bend and curve about them in such a way as to give the latter the appearance of eyes, as in the "augen gneiss." The amphibole is present in the tail-like appendages of the granulated phenocrysts, and also in the spaces between the sundered parts of the long drawn out crystals, and it is also mingled with the granules of the completely crushed phenocrysts. It often occurs in the cleavage cracks of the feldspar crystals. In its association with the mechanically deformed phenocrysts it shows no deformation in itself in being fractured or pulled apart.

Alteration of the amphibole. The amphibole alters or is replaced by two minerals, a variety of brown biotite, and a high doubly-refractive colorless mineral, which may be either calcite or dolomite. The biotite is perhaps the more abundant product although in some sections the colorless carbonate (calcite or dolomite) is also quite plentiful. The association of the colorless carbonate with the amphibole and biotite would seem to indicate that the carbonate contains magnesia and may be dolomite, rather than calcite. The biotite, where the genetic relations are definite, occurs at the frayed edges of the amphibole and also assumes a greater or less part of the original amphibole crystal. Sometimes the alteration of the amphibole to the carbonate is

through biotite, but often the change seems to be a direct one
to the carbonate. In those sections in which the biotite is most
abundant, there the carbonate is also most plentiful and the
amphibole of less importance. In most of the sections the amphibole is fresh and the biotite is usually absent, so that by far
the most abundant dark mineral in the gneiss is the former mineral. The biotite usually has a habit and orientiation like the
amphibole. Sometimes the biotite occurs in blades or bundle-
like aggregates of blades.

The carbonate sometimes wholly but usually only partly re-
places the amphibole. It also occurs as an infiltration between
the grains of quartz and feldspar. It is also at the frayed ends
of the prisms of both amphibole and biotite. In the replace-
ment of the amphibole by biotite and the carbonate, the end
product differs essentially in composition from the amphibole.
This must necessitate a considerable exchange in mineral substance which is interchanged by the agent of water solution.

Magnetite. This iron oxide is quite abundant in many of
the thin sections. It occurs in aggregates and streaks throughout the groundmass.

Apatite. This mineral occurs sparingly. It has the usual
habit of the long, needle-like crystals, exhibiting the common
transverse jointing.

Garnet. There is present in many of the thin sections a variable amount of the colorless garnet, grossularite. It occurs in
granular aggregates, in vein-like areas, and also as single individuals in the groundmass. It is almost colorless with a tinge
of reddish brown, and has a high index of refraction.

Zircon. This mineral occurs to a very small extent in the
groundmass. It also occurs as inclusions in the more weathered
phenocrysts of feldspar.

The texture of the groundmass. The parallel arrangement
of many of the constituents of the groundmass in a plane parallel to the longer axes of the mashed phenocrysts has already
been alluded to. In Pl. VII., fig. 2, is shown a thin section of
the gneiss in ordinary light. In this figure is shown a good
sized phenocryst of feldspar adjacent to which the minute dark

crystals of amphibole bend in curving lines. Farther away from the phenocryst the amphiboles lie in nearly straight lines. In many of the thin sections, as already stated, it can be seen that the streaky arrangement of the quartz and feldspar of the groundmass is due to the difference in the size of grains and to the arrangement of their slightly longer axes in a common direction. This laminated arrangement of the constituents of the Berlin gneiss gives it the gneissic texture, an appearance resembling that of the simple fluxion texture of a volcanic rock.

The recrystallization of the groundmass. The effect of pressure upon the feldspar phenocrysts is shown in their phenomena of granulation, separated and pulled out parts, fracturing and strain shadows. It is believed[1] that the pressure that produced the deformation of the phenocrysts in a similar way produced the gneissic texture of the groundmass under such conditions that the constituents of the latter recrystallized and assumed in general a parallel orientation.

The amphibole is thought to be secondary for the following reasons: This mineral not only occurs in the general groundmass, but is also formed with a similar habit in the appendages to the phenocrysts, which owe their production to the process of mashing of the latter. The amphibole occurs in the areas of partly and wholly granulated phenocrysts, and must have been developed during the process of the deformation. The parallel arrangement of the amphibole crystals, like the similar arrangement of the mica and amphibole constituent in many of the crystalline schists, is apparently due, as explained in the latter, to the development of such constituents while under great pressure and at a time when the rock was permeated with water solutions under conditions of at least moderately high temperature.

The quartz and feldspar of the groundmass likewise appear to be a secondary crystallization. These minerals are found filling the fractures in the pulled out phenocrysts, where they have the same character as in the general groundmass. The granular association of the quartz and feldspar is more like the texture of

---

[1] See Metamorphism and Rock Flowage by C. R. Van Hise: Bull. Geol. Soc. Am., Vol. 9, pp. 269-328. 1898.

the metamorphic schists or like the granular quartz in the jasper slates, and essentially differs from the complex textures of porphyritic volcanic rocks. Although there is not very marked difference in the diameter of the quartz and feldspar, yet their longer axes appear to lie in a parallel direction. The quartz and feldspar possess undulatory extinction only to a very small extent, and like the amphibole never show any fracturing. The constituents of the groundmass are thus in marked contrast with the extreme mechanical deformation of the phenocrysts.

From these facts, therefore, the conclusion is drawn that the groundmass is a development of recrystallization under pressure. That is, the pressure that has been effective in granulating the phenocrysts, has been sufficient, with its accompanying heat and water solutions, to cause the recrystallization of the groundmass.

Thus it appears that the gneissic structure as it now stands is only in part of mechanical origin. It is a combination of mashing and recrystallization, in which the coarser porphyritic constituent is granulated into smaller individuals and the groundmass recrystallized into larger granules. This process could under these conditions be prolonged until, in a general way, the grain of the rock would approach uniformity.

Such an explanation of the origin of the groundmass of the Berlin rock is in harmony with the conclusions of various investigators in the crystalline schists and other metamorphic rocks. In the sheared gabbros of the Lizard, Cornwall, Teall[1] describes a perfect foliation that has been induced by pressure. Teall says, "There is no reason to believe that foliation of the kind referred to in this communication can take place without molecular rearrangement."

Calloway[2] in studying the processes of metamorphism in the Malvern Crystallines is led to the following conclusions: "All the crystallines of Malvern were originally massive and amorphous. By a process of crushing and shearing, accompanied by the evolution of heat and the generation of intense chemical energy, a parallel structure has been here and there produced,

---

[1] Geol. Mag., November, 1886, p. 481.
[2] Proceedings of the Liverpool Geological Society, 1895–6, p. 454.

new minerals have been formed, and reconstruction on a large scale has been effected."

Van Hise,[1] in discussing the general origin of crystalline schists, concludes as follows: "The development of the crystalline schists is therefore believed to be explained as a process of chemical reaction induced by mechanical action, resulting in the constant solution and recrystallization of the material so as to accommodate it to the changed form."

III. THE WAUSHARA GRANITE.

GEOLOGY.

Distribution of the granite. The various outcrops of granite here described are located in the townships of Marion and Warren in the southeastern part of Waushara county. See map, Pl. 2. Their locality is about twelve miles northwest of the Berlin outcrop, and about thirty miles northwest of the area of rhyolite at Utley. All these granite areas are well exposed and range from small elevations to ridges and knobs fifty feet above the surrounding area. They are but a few miles apart. The intervening country is overlain by the Upper Cambrian sandstone and the Glacial Drift. Other areas of this granite will probably be found in the adjoining townships, but their description and location are necessarily deferred to a later report of the Survey.

The outcrops[2] to be mentioned, commencing the enumeration from the north, are as follows:

(1) The Berlin Granite Co.'s quarry, a small area located at the south quarter post of Sec. 8, T. 18 N., R. 12 E., in the township of Warren, Waushara county.

(2) Milwaukee Granite Co.'s quarry and outcrops on adjoin-

---

[1] Bull. Geol. Soc. Am., Vol. 9, p. 305.

[2] The writer is indebted to Mr. A. T. Lincoln for specimens from a number of the granite ledges.

ing farm of J. Macholl, making a broken ridge of granite something over a mile in length and about one-fourth of a mile in
width.   This ridge lies in the south one-half of Sec. 18, T. 18
N., R. 12 E., township of Warren, and in the southeast corner
of Sec. 13, T. 18 N., R. 11 E., in the township of Marion, Waushara county.   There is also an outcrop adjacent to this in the
southeast corner of Sec. 19, T. 18 N., R. 12 E.

(3) The several outcrops on the farms of Mr. Morrisey and
Mr Sivert in the eastern part of Sec. 27, T. 18 N., R. 11 E., in
the township of Marion.

(4) A large hill or ridge of the granite on the farm of Mr.
Scoby in the north one-half of Sec. 2, T. 17 N., R. 11 E., in the
township of Seneca, Green Lake county.   This outcrop has
been referred[1] to as the Seneca (Pine Bluff) Quartz Porphyry.
However, it is a typical granite and is in all respects like the
other granite ledges above enumerated.

Type of rock.   The rock constituting these areas is essentially
the same throughout.   It is a fine grained dull red granite consisting of feldspar and quartz, and a very small amount of fine
mica.   Differences in size of grain and color, due to processes
of weathering and granulation, are evident in some of the outcrops, but need no special mention.   The outcrops, especially
the larger areas, show numerous fractures and joints.   Dynamic
metamorphism has been effective in producing a cleavage sufficient to allow some of the granite areas to be readily quarried.
Paving blocks are an important product, and the granite is also
used for building and monumental purposes.

Dikes, veins, and fissures in the granite.   Many of the outcrops are cut by dikes of a black basic rock.   In the quarry of
the Berlin Granite Co. are two dikes intruding into the granite.
One of these is about a foot in thickness and strikes N. 70° E.
and dips 90°, and the other strikes slightly N. of E. and also has
a vertical dip.   The readiest cleavage of the granite at this
quarry is approximately parallel to these dikes.   Besides the
basic dikes there are several veins of fine granite that cut the
normal granite.   These granite veins are of the nature of the

---

[1] Geol. of Wis., Vol. II., p. 520.

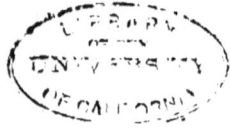

general type of granite. Numerous veins of quartz from a fraction of an inch to six inches in thickness extend in several directions.

In the area of the Milwaukee Co.'s quarry is a prominent vein of quartz varying in thickness from three to twelve inches that extends through the outcrop for a considerable distance. A large and persistent fissure containing much shattered rock also occurs in this quarry. The fissure runs in the direction N. 75° W. About one-fourth of a mile south of the Milwaukee quarry the wagon road crosses a low small ridge of granite, which is intruded by a large dike of the greenstone. This dike has a thickness varying from one to four feet, and strikes N. 85° E. The dike was followed the whole length of the granite ridge, a distance of about three hundred steps. The dike does not continue in a straight line for any distance, but cuts across and follows along parallel to fissures, apparently following planes of weakness in the older granite.

In the area of granite in the northwest corner of the town of Seneca, Green Lake county, there is a coarse phase of the granite. There is a basic dike at this place also, which is a very extensive one, and parts of it have been mashed into a mica schist.

MICROSCOPICAL PETROGRAPHY.

The thin sections show the rock to be a typical granite made up almost wholly of feldspar and quartz. Biotite occurs to a very small extent, and a few other minerals of secondary origin. The size of the grain is variable, dependent upon the amount of granulation of the quartz and feldspar in the different areas. In the study of the granite, under the microscope, the main interest lies in the comparison of its mineral composition and metamorphism with its consanguineous surface representatives, the Utley metarhyolite and Berlin rhyolite-gneiss. For this purpose our attention is necessarily directed to the metamorphism of the feldspathic constituent, although the quartz also shows definite metamorphism.

The quartz. The quartz appears to be of smaller grain than the feldspar, but whether or not this is entirely due to deformation it is not certain. Both the large and small crystals of quartz as a rule are characterized by strain shadows and sometimes fractures. It occurs as granophyric intergrowths with the plagioclase, and as stringers and web-like veins through the latter. In the latter case at least, their association with the feldspar indicates an origin due to the processes of mashing and lengthening of the feldspar crystals. The smaller grains of quartz often have a granular association with feldspar, like the groundmass of the rhyolite-gneiss.

The feldspar. The feldspathic constituent not only shows much evidence of mashing and mechanical deformation, but is also characterized by microperthitic growths and microcline twinning. The feldspar is of three types, plagioclase, microperthite, and soda microline.

The plagioclase is not only characterized by the usual polysynthetic twins, but in general throughout the sections it has a color slightly different from the microperthite and soda microline, having more of a yellowish tinge. It is usually studded with alteration products, and the twin lamellæ are often bent and broken apart.

About one-half of the feldspar is microperthite of the same general appearance as that occurring in the Utley rhyolite, and like that which also occurs to a very small extent in the rhyolite-gneiss. Several of the crystals that were cut through the plane of the brachy-pinacoid ($\infty P \check{\infty}$) were measured for the extinction angles of the perthitic growths. The angle approximated from +5° to +7° for one variety, and +17° to +21° for the other, indicating oligoclase-andesine and albite respectively. This indicates a composition of the feldspar in agreement with the microperthite of the other areas. It has been impossible to demonstrate in any individual case that the perthite in the granite is a secondary alteration of the simple plagioclase, as is the case in the Utley rhyolite. But from its close analogy with the Utley microperthite it is very probable that they have a common origin in both cases. The microperthite is fresh in appearance, and in this is in marked contrast with the plagioclase which usually contains numerous inclusions. The perthitic growths sometimes give evidence of some genetic relation to deformation, since the secondary feldspar often fills the spaces in the elongated and mashed plagioclase.

The cross twinning of microline is common to much of the feldspar. The cross lamellæ are well outlined, and the feldspar has the structure of typical microcline. It can be shown in several individual cases that the soda-microcline has originated from the simple plagioclase by the development of twinning on the application of pressure. The effects of pressure in bending the twin-lamellæ of the plagioclase and pulling the crystal apart and in inducing strain shadows in both the quartz and feldspar, is revealed in all the thin sections of the granite. The relation of

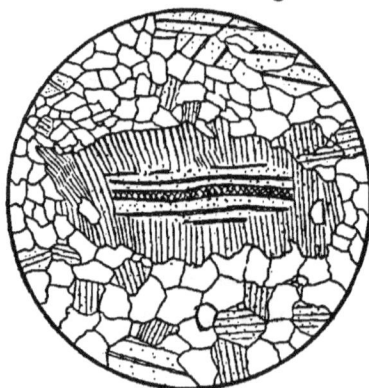

Fig. 13.—Section of granite showing granulation and development of secondary twinning in plagioclase.

the microcline to the plagioclase can best be brought out by a reference to the following figures:

In fig. 13 is shown a crystal of the plagioclase, the outer part of which consists of somewhat bent twinning lamellæ. The inner core consists of plagioclase twinned according to the albite law. On the application of pressure the outer boundary was twinned according to the pericline law, the pericline lamellæ dying out towards the center of the crystal. This pressure also produced granulation to some extent, and mashed and pinched the end of the plagioclase. In the process the albite lamellæ were somewhat bent. The inner core is appreciably more altered than the rim.

The alteration of plagioclase to microcline is shown in Pl. X., fig. 1. There is shown in the figure a well defined crystal of feldspar, the lower half of which is singly striated, while the upper half has the double twinning of microcline. The boundary between the two is distinct but jagged, and the microcline appears to have eaten its way into the plagioclase. The original crystal is plagioclase, twinned according to the albite law. In the processes of the deformation of the granite, cross twinning was probably induced, both sets of the new twinning lamellæ being finer than the original, and one set coinciding with the earlier twins. Narrow reticulating areas of untwinned feldspar ramify through the microcline. The plagioclase has a somewhat different color from the microcline, having more of a yellowish tinge, while the latter has a translucent aspect. In the upper right hand part of the crystal are two small remnants of the plagioclase within the area of microcline. The plagioclase contains numerous areas richly studded with decomposition products, while the microcline in general is free from such inclusions.

The relative abundance of alteration products in the plagioclase and microcline is better shown in Pl. X., fig. 2. In this figure is shown a crystal of feldspar made up of two parts; one part is studded with inclusions and the other part is microcline partly surrounding and ramifying through the former. The microcline is fresh and unaltered, and in marked contrast with the clouded area that contains many of the ordinary alteration products of feldspar, namely, sericite, chlorite, and microscopic colorless inclusions. At the upper left hand corner of the crystal the microcline and clouded plagioclase clearly grade into one another. Granulation has gone on to some extent, and a fracture extends across near the upper boundary of the lower band of microcline. The clouded part has indistinct twinning striæ that run nearly vertical in the figure, and coincide with one set of twinning lamellæ in the microcline.

Reviewing the phenomena presented in fig. 13 and Pl. X., figs. 1 and 2, the history of the feldspar is thought to be as follows: The original feldspar is plagioclase in good sized crystals, a soda-lime variety containing some potash, as indicated by the analysis of the granite and also by its similarity to the Utley feldspar, analysis of which is given on page 2. In the process of the deformation of the granite the plagioclase was in part granulated, and in part changed to soda-microcline by the devel· opment of secondary twinning.

The fact that the older plagioclase contains many secondary inclusions and the microcline is clear, might indicate that the latter differs somewhat in composition from the original plagioclase. Certainly there is a change in molecular arrangement, and under these conditions chemical change would be favored. But it is also true that secondary twinning can be induced in individual crystals without any attendant chemical change. A change in the potash constituent may be suggested. The presence of the sericite in the plagioclase, however, would seem to indicate that there is potash in the latter mineral, but its absence from the microcline cannot indicate that the latter bears no potash. Where the reverse conditions are present, as in the case cited by Whittle,[1] in which the plagioclase is secondary after microcline, and the microcline is clouded with sericite and the plagioclase clear, Whittle is inclined to believe that the absence of the sericite in the plagioclase is to be explained by the loss of the potash molecule in the transition to plagioclase. Although the example given in Pl. X., fig. 2, and that given by Whittle differ, as above indicated, they are alike in that the secondary mineral in both cases is fresh and the older feldspar is studded with the secondary sericite.

It may be that the production of sericite and microcline from the older plagioclase went on at the same time, since both are adjuncts to mechanical deformation. If the alteration of the plagioclase to sericite is subsequent to the mechanical deformation, then the difference in alteration may be due to unlike molecular arrangement of the secondary microcline and original plagioclase. The persistent cloudiness and alteration of the plagioclase in contrast with the fresh microcline and microperthite is a general phenomenon throughout the granite area. In the Utley rhyolite the older cores of plagioclase seem to have

---

[1] Bull. G. S. A., Vol. 4, 1893, pp. 162-4.

been altered previous to the development of the clear microperthite of the rims. The feldspar phenocrysts of the Berlin gneiss are usually very free from inclusions, and no traces of the ordinary twinned plagioclase are present.

Granophyric structure in feldspar and quartz. Granophyric structure of quartz and feldspar is common in many of the larger crystals of feldspar. In its occurrence it seems to have its origin in a process somewhat allied to granulation. Some of the crystals of soda-microcline contain numerous streaks of quartz that seem to be along shearing planes or tensile fractures in the feldspar. The streaks of quartz in the feldspar usually have the same orientation. Seen in cross section the ends of these streaks may or may not be separated by areas of soda-microcline. In the latter case they appear as curving canals and polygonal outlines like the usual figures of granophyre. Its association with the secondary soda-microcline and streaky character in this mechanically deformed rock would seem to indicate its secondary origin.

Other minerals of the granite. Biotite occurs to a small extent in the granite, in small crystals and in veins and spaces between the quartz and feldspar. Some magnetite and other iron oxide is also present. The sericite and small amount of chlorite have already been mentioned in connection with the alteration of the plagioclase.

Composition of the basic dikes. The greenstone dikes that penetrate the various outcrops of granite are usually rather massive. Under the microscope the principal mineral is seen to be a green variety of amphibole in very small, well outlined crystals. There is some variation in the mineral composition. In some localities the plagioclase is entirely absent, while in other places it is abundant. In some cases brown biotite constitutes a great proportion of the dark mineral. The amphibole is a green variety and is probably common hornblende. The rock of the dikes probably should be classed as a metamorphosed diorite or peridotite.

SUMMARY.

(1) The three areas of igneous rock described in the foregoing are of pre-Cambrian age, and appear to have been islands off the ancient shore line of the large and continuous pre-Cambrian area of the northern part of the state. The topography of these areas, with others of the Fox River Valley, indicates that the region was not reduced to a base level at the time of the encroachment of the Paleozoic sea.

(2) The rocks of these isolated outliers are congenitally related, as shown by the close agreement in chemical composition shown by analysis. The similarity in composition of these outcrops with that at Baraboo establishes the fact that a petrographic province of considerable extent is represented in these pre-Cambrian formations.

(3) These congenital rocks represent various phases of the parent magma, and range from volcanic flows to masses of deep seated origin, with corresponding textures.

(4) There is a difference in the metamorphism in the various outcrops due to the unlike conditions through which they have passed since their original crystallization. In the Utley metarhyolite the metamorphism has taken place under static conditions, no cleavage is developed in the rock, and alteration of the phenocrysts and groundmass has taken place through chemical change without the aid of mechanical deformation. The Berlin rhyolite-gneiss, on the other hand, has been subjected to extreme deformation, the original rhyolite has been mashed into a gneiss, a perfect system of cleavage and fractures has been developed throughout the area, and chemical reaction and strains induced by mechanical action have resulted in the almost complete recrystallization of the rock. The metamorphism of the Waushara granite has been in part static and in part mechanical, cleavage has been developed and granulation of the quartz and feldspar has taken place to some extent.

(5) The original feldspar constituent of the various areas is a

plagioclase, a soda-lime-potash variety, and so far as could be determined, this species is common throughout the areas.

(6) The plagioclase shows stages of alteration from its original condition. This metamorphism takes place in two ways, by a development of microperthite through chemical change under static conditions, and by the development of soda-microcline by secondary twinning on the application of pressure. The characteristic change in the Utley metarhyolite is to microperthite, the change in the Berlin gneiss is to microperthite and to soda-microcline, and the Waushara granite shows both these products of alteration in about equal proportion.

(7) The nature of the microperthite is that of an interlocking growth of two species of feldspar, one species being albite with an extinction angle of 19° on the brachypinacoid ($\infty P \check{\infty}$), and the other species is oligoclase-andesine with an extinction angle of 5° to 7° on the same face. The perthitic growths tend to develop a definite crystal form, are bounded by domes, prisms, and pinacoids, and appear to have a regular arrangement in the crystals in which they occur.

The microperthite is secondary and has developed in three ways, each kind differing slightly from the other, yet all being brought about by similar processes of recrystallization. These processes have developed the microperthite as follows: (a) As secondary enlargements or outgrowths upon older crystals of plagioclase; (b) As secondary intergrowths or regenerations of older plagioclase; (c) As complete new phenocrysts in the groundmass. All three are exhibited in the Utley metarhyolite, while in the Waushara granite the perthite is probably wholly a regeneration of the plagioclase.

(a) In the process of growth by secondary enlargement the microperthite, Pl. VIII., figs. 1 and 2, as seen in cross section, occurs as rims which surround an inner core of plagioclase, the microperthitic growths having an orientation nearly coinciding with the plagioclase of the core. The latter is fractured, and contains alteration products, while the microperthite is free from fractures and is always fresh. In this secondary growth of new microperthite about the plagioclase, is an example of the enlargement of phenocrysts in an igneous rock which is analogous to the enlargement of the crystal grains in the metamorphic sedimentary rocks.

(b) As a regeneration of older plagioclase the perthitic growths have an orientation in common with the "mother crystal." The new perthitic growths tend to have definite crystal form (Pl. VII., fig. 1, and fig. 1, p. 16). The process of change from plagioclase to microperthite, *perthitization,* begins at the boundary of the crystals and along fractures, and the plagioclase is invariably clouded with alteration products while the microperthite has a fresh appearance. The new perthitic growths in a single crystal appear to have their longer axes in a common direction and this direction is in either of two planes. They may be developed with their longer axes in the plane parallel to the composition face of the albite twins, the brachy-pinacoid, or they may be developed in the plane parallel to the composition face of the pericline twins, the rhombic section. It is also believed that these new growths, in whatever plane their longer axes extend, have an orientation nearly coinciding with the older plagioclase, and that the process of perthitization is analogous to paramorphism. Various stages of the process of change to perthite are present, making the feldspar phenocrysts appear irregularly mottled and the perthitic growths with irregular boundary. This incomplete growth of the new feldspar species within the plagioclase, and the fact that the sections are usually cut obliquely to the brachy-pinacoid or rhombic section, gives the feldspar a mottled, irregular appearance under the crossed nicols and the perthitic growths then appear to possess no symmetry of structure.

(c) In the development of new phenocrysts of microperthite in the metarhyolite, the new growths often possess sutures from which the species of feldspar of slightly different orientation project in closely intergrown fibres. (Figs. 5 and 6.) They possess irregular finger-like shapes and seem to have developed along minute fractures in the groundmass.

(8) In the Berlin rhyolite-gneiss the feldspathic constituent shows much evidence of deformation in being fractured, pulled

apart, and granulated, and where the deformation is not too great the feldspar usually has the cross twinning of microcline. In the Utley metarhyolite, which is almost free from any evidence of deformation, the typical cross twinning of microcline was not observed. In the Waushara granite where the mashing process was not so advanced as in the rhyolite-gneiss, various stages of the change from plagioclase to soda-microcline were noted. (Pl. X., figs. 1 and 2.) The plagioclase is usually studded with sericite and biotite and is clouded with minute secondary inclusions, while the soda-microcline is translucent and has a fresh appearance. The secondary twinning lamellæ are finer and are superinduced (Pl. X., figs. 1 and 2) upon the earlier twinning lamellæ of the plagioclase.

APPENDIX.

SOME PREVIOUS WORK ON MICROPERTHITE AND SODA-MICROCLINE.

It is proposed to present at this place a review of some of the literature on the perthitic and microcline structure of feldspar, in order to show how much the work of previous investigators accord with one another, and with the conclusions arrived at in the present paper.

Perthite was first described by Thomson[1] as a new species of feldspar, but later it was shown by D. Gerhard[2] to consist of interlaminated orthoclase and albite. Gerhard made an analysis of this feldspar and showed that one set of lamellæ contained 12.16 parts of potash to 2.25 parts of soda, and that the other set of lamellæ contained 3.34 parts of potash and 8.50 of soda. Many occurrences of perthite have since been noted by various authors.

---

[1] Phil. Mag. Vol. 22, p. 189, 1843.
[2] Zeit. d. Geol. Gesell. Heft 4; Band 14. p. 151, 1862.

When the perthitic structure is discernible only with the microscope, according to Becke, it is called microperthite. Other names have also been applied descriptive of the composition and structure of the perthitic intergrowths, such as orthoclase-microperthite, microcline-microperthite, and cryptoperthite, which appear in many cases to show a close relation to soda-microcline and soda orthoclase.

H. Credner[1] describes the albite in perthitic intergrowths of orthoclase and albite, as follows:[2] "Die aus dem Perthitartigen Feldspath extrahirte Albitsubstanz siedelt sich in anfänglich kleinen, almälig wachsenden Krystallen und Krystallincrustaten entweder auf des Oberfläche, am Füsse oder in des weiteren Umgebung des Mutterminerals, in ersterem Falle in paralleler Stellung zu diesem an." The relative amounts of the potash silicate to the soda silicate, in this case was as 4 is to 1. It is apparently Credner's inference that none of the orthoclase received a molecular rearrangement in this perthitic intergrowth although the albite molecule has been extracted from it.

A. Sauer[3] has described similar alterations of the feldspar in granites of Socotra in which the original crystal is soda orthoclase, the relation of potash to soda being 9 to 5, with only a trace of lime. In this case the older crystal is monoclinic and the change to albite is along the edges of the soda orthoclase, and in streaks across the latter, the albite lamellæ being developed with their longer axes normal to M, but the line of growth being in a direction parallel to the M face. Two kinds of orthoclase were noted, one that was colored reddish, and another that was whitish and fresh in appearance like the new albite.

The whitish feldspar constituents were analyzed and showed an excess of albite over orthoclase substance. Of this Sauer says:[4] "Ob dieser Vorgang schliesslich zu einer vollständigen Verdrängung des letzeren durch das Natronsilicat führen könnte, ist theoretisch nicht unwahrscheinlich, an vorliegen-

---

[1] Zeit. d. deutsch. Geol. Gesell. Band 27, pp. 104-261, 1875.
[2] l. c. p. 213.
[3] Zeit. d. deutsch. Geol. Gesell. Band 40, pp. 146-152, 1888.
[4] l. c. p. 149.

den Material aber nicht mit Sicherheit zu constatiren. Es. scheint eher, als ob gleichzeitig mit der Albitisirung eine Art von Regenerirung reiner Orthoklas substanz stattfände."

J. Lehhman[1] has also described secondary growths of feldspar producing a perthitic structure in which he explains the growth of the albite as taking place "in fractures in orthoclase due to contraction which are formed parallel to the pinacoidal axis and to the vertical axis. These fractures which have become partly widened by etching solutions form the conducting canals for the penetrating albite." That is, that ready albite substance penetrates the existing cracks and heals them. Sauer refers to Lehmann and does not agree with him in the explanation of the way in which the perthitic structure originated secondarily, holding that it is not necessary that cracks due to contraction should be present for the growth of the albite, but that the growth of the latter is probably purely chemical from the older soda orthoclase.

W. C. Brögger,[2] in his work upon the syenite-pegmatitic dikes of Southern Norway has given an excellent description of the feldspar prevalent in that region. The feldspar there occurring are albite, microcline, microperthite, and cryptoperthite, the general appearance and relations of which are very similar to the feldspar described in the foregoing paper. The composition of the feldspar described by Brögger, however, is alkaline and almost free from lime, while that described in the present paper contains more lime than potash.

Brögger concludes[3] that there are two kinds of microperthitic intergrowths, the origin of one being secondary and of the other primary. That of secondary growth is developed in the plane of the vertical axis, the orthopinacoid, "durch eine Spaltung des ursprünglich auskrystallisirten Natronorthoklases"—"N i c h t secundär, sondern ursprünglich ist dagegen nach meiner Ansicht die äussert feine mikroperthitische Verwachsung von Albit und

---

[1] Jahresbericht d. Schles. Gesell. f. Vaterländische Cultur, Band 63, pp. 92–100, 1885.

[2] Zeit f. Kryst. Band 16, pp. 524–53; Brögger, pp. 520–564.

[3] l. c. p. 537.

normalem Kaliorthoklas im Feldspath vom Gomsö-Wege; diese primäre mikroperthitische Verwachsung unterscheidet sich in mehreren Beziehungen wesentlich von der soeben als secundär angenommenen in, erster Linie auch durch die verschiedene Lage der Verwachsungsebene selbst, welche bei dem Feldspath vom Gomsö-Wege (wie auch bei allen unten zu beschreibenden Mikroperthiten aus den Gängen der Gegend des Langesundfjords) einem steilen Orthodoma, wahrscheinlich ungefähr 8 $P\bar{\infty}$, parallel ist, während die Verwachsungsebene bei den secundären mikroperthitischen Verwachsungen das Orthopinakoid ist." Brögger also concludes that since the pure potash silicate $K_2Al_2Si_6O_{16}$ occurs as asymmetric microcline, that monosymmetric orthoclase may be considered as a cryptolamellar microcline, as held by Michel-Lévy, Mallard, Rosenbusch, Groth, and others, "eine Erklärung, deren Möglichkeit nicht bestritten werden kann."

Brögger assumes the existence of very fine interlamination of albite and orthoclase, not discernible under the microscope, which he terms "Kryptoperthit," and that there would be all gradations of this cryptoperthite into soda-microcline on the one hand, and coarser microperthite on the other.

As quoted above Brögger describes two kinds of perthitic intergrowths differing from one another in the position of their planes of growth, the position of one being in a plane parallel to the orthopinacoid $\infty P\bar{\infty}$ and of the other in a plane approximately parallel to a steep orthodome 8 $P\bar{\infty}$. It is not apparent however, upon what evidence he bases his conclusion that the former variety is of primary nature and the latter secondary.

N. V. Ussing[1] describes the alkali feldspar in the Greenland syenites, which occur as orthoclase-microperthite, microcline microperthite, cryptoperthite, soda-orthoclase, and soda-microcline. In general both silicates (x K AL Si_3O_8 + y Na Al Si_3O_8) are crystallized into mixture crystals, within each of which the vertical axis or prism plane is common for all the molecules. These mixture crystals are either homogeneous soda-orthoclase (soda-microcline) or heterogeneous perthite with subdivisions

---

[1] Meddelelser om Gronland. Heft. XIV., 1894. pp. 15–106.

microperthite and cryptoperthite. The former are united with the latter through transitions and can be considered as cryptoperthite in which the individual lamellæ of potash and soda feldspar sink into sub-microscopic dimensions.

According to Ussing the intergrowths may be primary or secondary. The intergrown planes between the two feldspars of the primary growths are of two kinds, those that arise between subsequent crystallizing parts of the two feldspars, and such as arise between simultaneous crystallizing parts. The intergrown planes of the first kind follow crystal planes, the planes of the other kind do not correspond to any crystallographic plane but owe their position to certain crystallographic elements like those apparent in pericline twins. In orthoclase and albite the planes follow along the steep dome $(80\bar{1})$; in cryptoperthite (microcline) they follow along the two pyramidal planes $(8\bar{6}1)$ and $(86\bar{1})$.

. Ussing also states that the boundary planes of the secondary perthitic lamellæ become the same as the primary under certain conditions, and under other conditions they do not, and that one must decide in each individual case how far they are primary or secondary.

From the foregoing it will be seen that the composition of the perthitic feldspars described, differs somewhat from one another, but in general they are almost wholly alkaline, lime entering into them in only a very small proportion. In contrast with these, those described from Wisconsin contain more lime than potash, the molecular proportions being, $Na_2O : Ca O : K_2O :: 11.61 : 9.86 : 5.43$.

The various authors mentioned above, and others not reviewed here, have noted the fact that the perthite lamellæ are developed along certain planes of the feldspar in growths considered either primary or secondary or both. These planes of growths "Verwachsungebene" are described as being parallel to the edge between P and M; nearly parallel to the base and vertical prisms; parallel to the orthopinacoid; parallel to the orthodom 8 P ∞; parallel to the pyramid planes 6 $P\frac{3}{4}$ $(8\bar{6}1)$ and 6 $P\frac{3}{4}$ $(86\bar{1})$. The growths as described are usually along the vertical axis, other planes of growth being not so common. While there are differ-ent planes of growth, as described by various authors, yet all have noted that there is in certain cases at least, an apparent uni-formity in the structure of the perthitic feldspar.

The conclusions arrived at by the author as to the nature and origin of the microperthite and soda microcline and described in this paper, are given in the summary, p. 55. The original feldspar is plagioclase, which has altered to microperthite and microcline. The alteration to microperthite is by a chemical change under static conditions, and to microcline by a molecular change on the application of pressure.

The microperthite is developed in three ways, of similar na-ture, by enlargement of older plagioclase, by regeneration of older plagioclase, and by entire new growths in the groundmass. The microperthite of the enlargements have an orientation de-pendent upon the crystal upon which they are grown. The mi-croperthite due to regeneration is believed to have its planes of growth either in the plane of the composition face of the albite twin ∞ P $\check{\infty}$ or in the plane parallel to the composition face of the pericline twins, the rhombic section. But in whatever planes the growths may have their longer axes developed, it is believed the new species of feldspar approximate as near as possible the crystallographic symmetry of the parent crystal from which they are derived, and that the regeneration of the plagioclase into mi-croperthite, *perthitization*, is very similar to processes of para-morphism and not unlike the process of schillerization as de-scribed by Judd.[1]

---

[1] Q. J. G. S., Vol. 41, 1885, pp. 374–389.

1

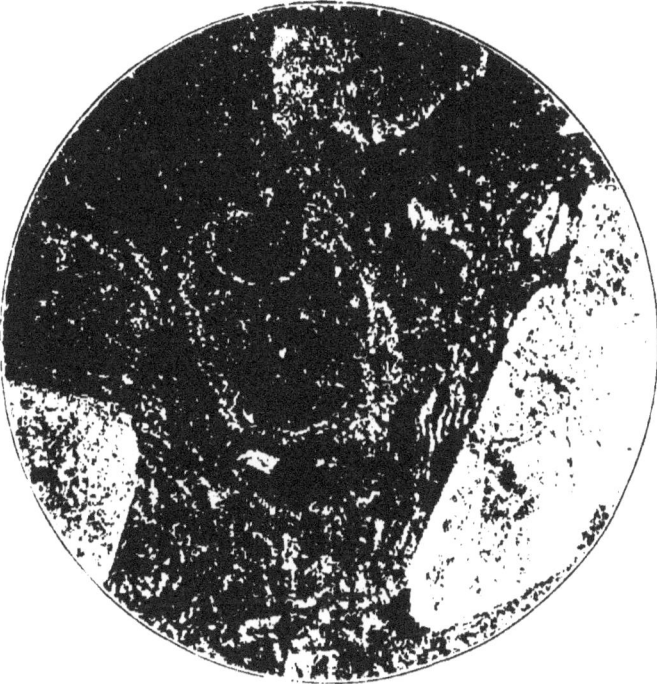

2

PLATE VI.

Figure 1.   Specimen of the metarhyolite from the layers or beds of spheroids.   Two-thirds natural size.

The weathering of the rock along the spheroidal partings produces an appearance, as seen in cross section, of a ring or chain-like structure on the somewhat weathered surfaces of the rock. As the rock becomes much weathered the spheroids stand out as nodules, having much the appearance of weathered spherulites, but differ from the latter in possessing no radial structure.

Figure 2.   Metarhyolite.   Microsection 3845.   Ordinary light, x 50.

The figure shows phenocrysts in a fine groundmass.   In the center of the figure is an example of perlitic parting.   The groundmass has recrystallized along the parting, producing a zone of parallel fibers of feldspar extending normal to the parting.   Attention is called to the similarity of this perlite to the spheroids shown in Pl. VI., fig. 1.   In the groundmass are numerous patches and curved strips of fibrous growths of feldspar.

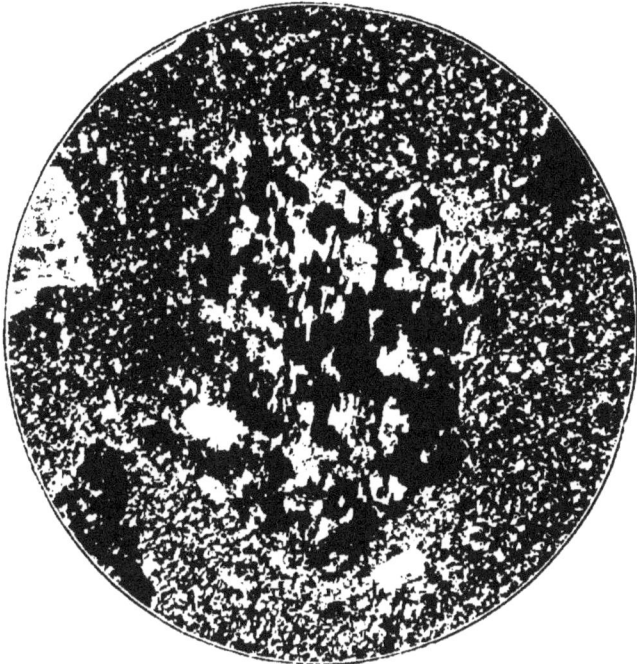

1

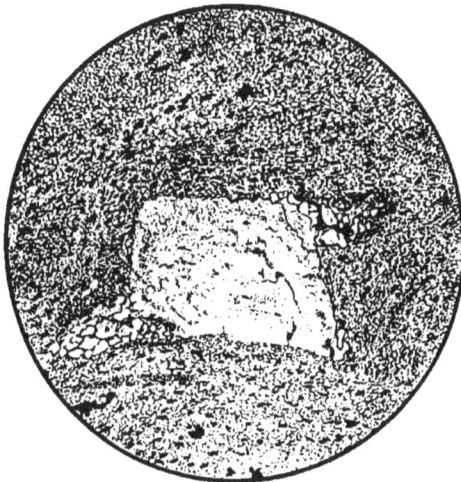

2

PLATE VII.

Figure 1. Metarhyolite. Microsection 3846. Polarized light, x 40.

In the center of the field is shown a typical microperthite crystal cut nearly parallel to $\infty P \bar{\infty}$ and showing basal cleavage. The light part is albite with an extinction angle of 19°; the dark part is oligoclase-andesine, Ab_3An_1, with an extinction angle of 3° to 6°. The microperthite in this case is a product of regeneration. The new growths are developed with an orientation approximately coincident with the original plagioclase, and the new growths tend to possess definite crystal forms.

Figure 2. Rhyolite-gneiss. Microsection 3816. Ordinary light, x 30.

The figure shows a good-sized phenocryst of feldspar, around which the small dark crystals of amphibole bend in curving lines. Farther away from the phenocryst the amphiboles lie in nearly straight lines. Tail-like appendages extend from the corners of the phenocryst parallel with the lines of amphibole crystals.

1

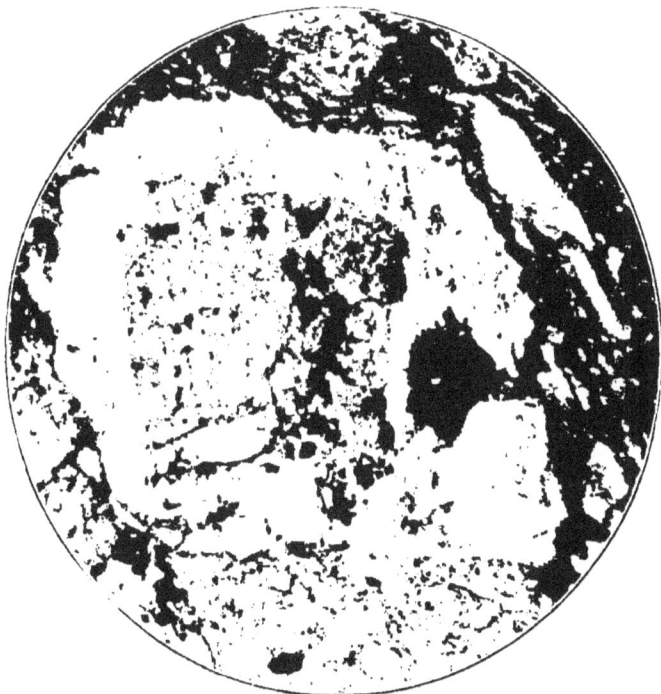

2

PLATE VIII.

Figure 1. Metarhyolite. Microsection 3845a. ' Ordinary light, x 60.

The figure shows a feldspar phenocryst made up of an inner core of original plagioclase surrounded by a thick margin of secondary microperthite. The inner core of twinned plagioclase contains fractures which do not penetrate the rim. These fractures, and the plagioclase as a whole contain numerous decomposition products. The microperthite of the margin contains no fractures, and is free from alteration products. The sharp boundary between the original core of plagioclase and secondary rim of microperthite is to be noted. The perthitic growths of the enlargement have their longer axes parallel to the $\bar{c}$ axis of the core. In the lower left hand corner of the plagioclase core is a small patch of microperthite lying between the lower fracture and the lower end of the core, which is due to the regeneration of the plagioclase.

Figure 2. Metarhyolite. Microsection 3845a. Ordinary light, x 60.

The feldspar phenocryst is made up of a core of plagioclase with secondary enlargement of microperthite, the boundary between the core and rim being very clear. The core contains fractures which do not penetrate the rim, and these fractures and the core as a whole contain much sericite, biotite, iron oxide and other decomposition products. The microperthite contains no fractures, and is almost free from decomposition products. In the right side of the rim is a black crystal of iron pyrite, around which the microperthite has not wholly succeeded in growing.

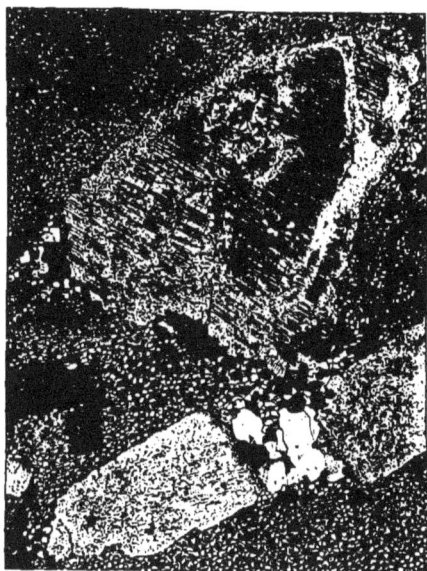

1

2

3

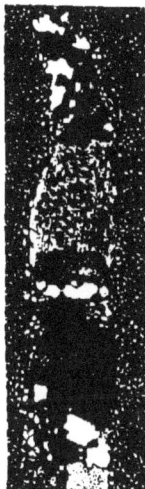

4

PLATE IX.

Figure 1. Rhyolite-gneiss. Microsection, 3817. Polarized light, x 20.

The figure shows two crystals, the one above having been pressed down upon the one below and broken it apart. The upper crystal is fractured and granulated near the middle part, and it shows the fine cross-twinning of typical microcline. It contains streaks and irregular areas in which the cross twinning is not developed. The lower crystal has been pressed apart and the intervening space is filled with quartz, feldspar, and blue amphibole. Tail-like appendages project from the upper corners of the upper crystal in a direction parallel to the longer axis of the lower crystal, which is the plane of cleavage.

Figure 2. Rhyolite-gneiss. Microsection, 3815. Polarized light, x 35.

Phenocryst of feldspar showing fractures and strain shadows.

Figure 3. Rhyolite-gneiss. Microsection, 3809. Polarized light, x 20.

The figure shows a part of a mashed and granulated feldspar crystal. The space in the center of the figure between the ends of the phenocryst is made up of the broken parts of the phenocryst and secondary quartz and feldspar.

Figure 4. Rhyolite-gneiss. Microsection, 3829. Polarized light, x 16.

Part of a phenocryst of feldspar showing process of lengthening by fracturing, granulation and recrystallization.

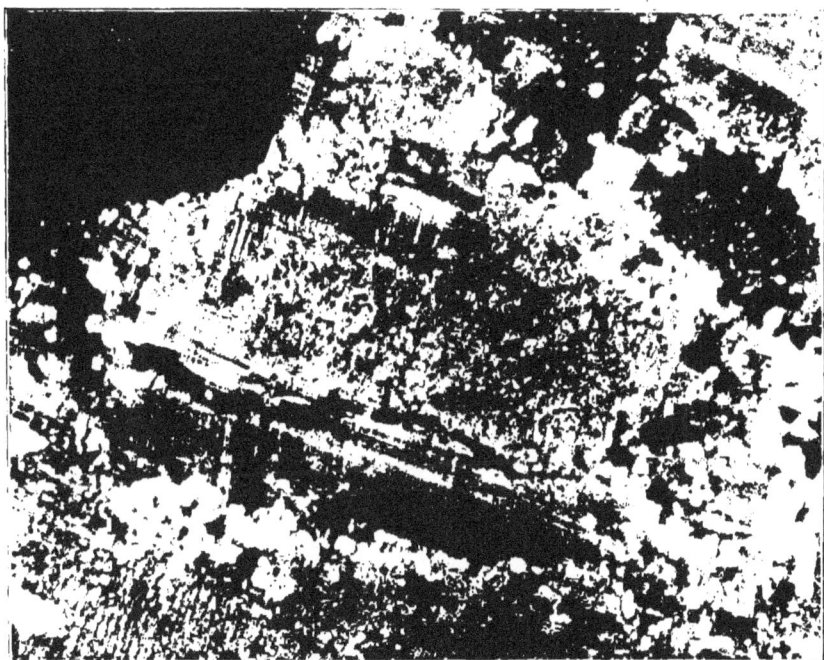

PLATE X.

Figure 1. Granite. Microsection, 4385. Polarized light, x 40.

The alteration of plagioclase to soda-microcline is here very clearly shown. As indicated by the irregular boundary between the two parts the microcline has the appearance of having eaten its way into the plagioclase. This molecular change is brought about by pressure, and one set of the new and finer twinning lamellæ coincides with the earlier twins.

Figure 2. Granite. Microsection, 4393. Polarized light, x 50.

Almost the whole of the figure is taken up by what was once a single plagioclase crystal. In the lower left hand corner of the figure granophyric structure has been produced by the fracturing of the plagioclase and the formation of canals of quartz. The microcline clearly grades into the clouded plagioclase. Traces of the twinning lamellæ still remain in the plagioclase, and one set of the twinning lamellæ of the secondary soda-microcline coincides with the earlier twinning plane.